REVEALED BIODIVERSITY

An Economic History of the Human Impact

REVEALED BIODIVERSITY

An Economic History of the Human Impact

Eric L. Jones

Emeritus Professor, La Trobe University
and
former Professorial Fellow Melbourne Business School
University of Melbourne

NEW JERSEY • LONDON • SINGAPORE • BEIJING • SHANGHAI • HONG KONG • TAIPEI • CHENNAI

Published by

World Scientific Publishing Co. Pte. Ltd.
5 Toh Tuck Link, Singapore 596224
USA office: 27 Warren Street, Suite 401-402, Hackensack, NJ 07601
UK office: 57 Shelton Street, Covent Garden, London WC2H 9HE

Library of Congress Cataloging-in-Publication Data
Jones, E. L. (Eric Lionel)
Revealed biodiversity : an economic history of the human impact / by Eric L Jones.
p. cm.
ISBN 978-9814522564 (hardcover)
1. Agriculture--Environmental aspects--History. 2. Agriculture--Economic aspects--History.
3. Nature--Effect of human beings on--History. 4. Human ecology--History. I. Title.
HD1433.J66 2014
333.95--dc23

2013019823

British Library Cataloguing-in-Publication Data
A catalogue record for this book is available from the British Library.

In-house Editors: Philly Lim/Dipasri Sardar

Typeset by Stallion Press
Email: enquiries@stallinpress.com

Printed in Singapore

For Cliff Irish
And to the memories of Bill Dreghorn and Colin Tubbs
Naturalist companions of my schooldays

CONTENTS

PREFACE

Revealed Biodiversity takes an economic historian's view of the environment. The reason for this is that the natural world, as we experience it, is substantially a product of the economy and approaching it from this angle raises questions that are ordinarily easy to miss. Nature is not pure, abstract, independent of humanity, nor a mere punch bag for our greed. If we took nothing from the natural world we would drain it of interest to humanity; it would be a painted backcloth. Instead, first, we shape the environment through economic decisions; secondly, the decisions can be beneficial as well as harmful, either outcome being of academic and scientific interest; and thirdly we impart meaning to nature by consuming it. All the processes can be charted historically and many of the episodes will appear in the course of this book. Alongside the substance of the text runs an examination of the way in which environmentalists treat, or too often do not treat, the historical issues.

The environment is not simply a damaged remnant left over from human exploitation. Uses are not automatically threats; sometimes they broaden and increase natural variety or build harmonious landscapes in which both man and beast flourish. Despite this, almost everything written today about human dealings with the environment is negative. It is not Pollyanna-ish to lean a little to the opposite side, towards balancing the books. In many respects the environmental situation is bad, even frightening, but squinting at the world through a single eye can produce only misunderstanding and lead to bad policies. Let us not lament ecological losses, as if loss is all there is: let us be fair and take account too of the charm of humanised landscapes and the improvements in our ability to enjoy nature. It is these that create 'revealed biodiversity' — the biodiversity we see, or looked at from the economist's viewpoint, consume.

Three features are overlooked in much of the environmental literature and this book will try to provide them. Let us take them in order of importance: first, the social gains, the rise in human population, the lengthening of life expectancy, and the better standard of living, are often set at naught compared with reduced numbers and distributions of some, indeed many, birds, animals, butterflies and flowers. No one deplores the losses more than me, as a lifelong naturalist. Nor does anyone resent the fact of changes more than I do, given that in the developed world they have often been unnecessary (bad policies again). Yet omitting the positive side of the environmental balance sheet is an inhumane form of accountancy, reckoning nothing for the well-being of poor people or economic growth in less-developed countries.

Secondly, the offsetting gains of human intervention, such as the recoveries of threatened species or land being taken out of agricultural production and (inadvertently) returned to nature on a large scale, are commonly skated over, ignored, or denied. Better to report these trends than imply that humanity has never ceased pushing the natural world to hell in a handcart. Much of this book consists of histories of fluctuations in wildlife which run counter to the narrative of uniform decline.

Thirdly, where ecosystems do receive a measure of conservation, the case for rescuing nature is sometimes advanced separately from claims for the conservation of the landscape on historical grounds. Naturalists tend to ignore cultural history or actively disdain it. The point may seem minor and relatively speaking it is. But the attitude is needless and undermines the coalitions of naturalists and landscape historians most likely to win campaigns for conservation. Moreover it is an example of blindness to human history on the part of environmentalists. Because this is more thoughtless than antagonistic (though it is that on occasion), the breach can and should be healed.

The position adopted here is plainly that of an economist and historian. The book is after all an economic history of the environment. Despite an absolute minimum of the terminology of economics and none of the techniques, biologists and conservationists will find the stance unfamiliar and may find it uncomfortable. My hope is that it may nevertheless help them to put names to the anti-market, command-and-control, attitudes that pervade conservation, and see that these are not the only ways of shining light on the subject. Ecology is rather politicised, more so than other sciences except for climatology. But again there is an element of thoughtlessness rather than malice in this and the issue can be teased out instead of leaving anti-market ideology uncontested. The economic calculus is relevant: for instance, wherever the environment has been protected, people must have been willing to pay — the market must have intruded. There is an opportunity cost, since the money might have bought something else. Society has plenty of other urgent needs.

Academic and intellectual life have become so professionalised and compartmentalised — the fashionable metaphor is of separate silos — that ranging across two or three subjects, never common practice, is even less common than it was. The broad approach of the economic historian, joining economics and history, brings to the fore developments in the natural world normally left out of account or played down. New relationships come into sight from this perspective. Once in a while I also throw in an observation based on my own experience and I hope this too proves illuminating. A personal touch is out of fashion, in favour of academic writing that seems to me dry and implausibly detached. The purpose here is to add to the argument, to humanise it, and avoid the artificial pose of detachment.

Among environmental history's problems is a tendency to hark back to a Golden Age when nature was supposedly unsullied. At least this is better than comparing later times with no previous baseline at all. The baseline issue is serious: a reference point is needed against which to evaluate changes, though the usual practice is to criticise deviations from a Garden of Eden where, so to speak, the grass was always greener. Not only are lost species bewailed but new arrivals are condemned as though they must be unnatural — despite the fact that they present such fascinating experiments. This lachrymose attitude sinks into a larger mind-set in which the spread of people around the globe and the development of their economies are labelled destructive and wrong. Nine times out of ten, rank prejudice is smuggled into ecology through this back door, since the people slated for bringing about change are almost exclusively the Europeans.

Opposing the European record, shorn of all praise for its human achievements, romantic portraits are offered of pre-European ecologies and the role in them of Noble Savages. These suppositions, which form the background of much work in the field even when they are not the foreground, are simply unhistorical. They get in the way of developing global environmental history, in which the changes brought about when non-European populations exploded into fresh territory ought to figure alongside Europe's expansion. The present book attempts no complete history in this respect but does set against the spread of Europeans (whom Walter Bagehot, writing at their peak, called the 'conquering *swarm*') a description of the closest modern equivalent, the global impact made today by East Asians.

Economic growth in East Asia has provided enough wealth to convert biological items that were once luxuries into commodities affordable and relentlessly sought on behalf of vastly more members of Asian society. There are signs that this is starting to calm down, as the tentative shoots of 'Green' movements begin among the Chinese and Overseas Chinese. But for traditions to alter completely may take generations. In the meantime accounts of the East Asian impact are necessary to offset one-sided histories in which the villains are always Europeans. (This was always nonsense, since Palaeolithic peoples in the Pacific islands had exterminated 20% of the world's avifauna before ever the Europeans arrived.) Two, or even three, wrongs do not of course make an ecological right: what is needed is a more dispassionate and complete history.

Closer to the ground, as it were, is the choice of baseline when particular sets of ecological changes are discussed. In much of the literature aimed at a wide readership, and hence significant for politics and policy formation, there is a tendency to concentrate on species and places that may be said to be declining. Recoveries of threatened species are reported (the case of the peregrine falcon comes immediately to mind) but happy events make up little of the news overall. Too often the results are negatively biased accounts of trends in which species that gain aggressively from habitat change are dismissed chauvinistically as interlopers.

Attention inclines to centre on the dwindling numbers of 'rare' species which are thought to have been common in the past — as if ecosystems have been stable through time. This assumption usually takes some rather arbitrary period as its baseline, perhaps as a nostalgic tribute to when the observer was young or to the time when detailed records first begin. Subsequent history is set against these starting points, which can mislead through failing to recognize where they stand amidst the fluctuations produced by agricultural and other changes. A difficulty is that during former centuries, even in the twentieth century, natural history observations were often sporadic rather than systematic. Nevertheless, with careful searching information may be found and even records that seem merely indicative can act as correctives to the thought that everything has invariably deteriorated. Changes in the bird and butterfly populations of southern England are used here to illustrate how dangerous it can be to set the baseline during the agricultural (or arable) depression of the late nineteenth century. At that period falling investment in farming and forestry created what naturalists conceived as superb habitats. They were brought back into production during the twentieth century, reducing again the populations of species that had been plentiful during the depression years, and causing naturalists of the philatelic or list-making persuasion to scour the countryside for them in the belief that they cannot really have disappeared.

Understanding the degree to which the environment is an economic artefact is clarifying. A sketch of the history of economic development will help to substantiate this. Agricultural history is especially relevant because farming uses one factor of production, land, so heavily. Farming is ecologically significant, that much is obvious. Related to agricultural history are the histories of hunting, shooting and angling, all of which are intimately bound up with ownership or access to the land. They were enormous in scope and taken to be the birth-right of every settler of wild land, besides conferring status on participants in the old-settled countries. Status continues to be associated with shooting, especially in England. Scrupulously omitting to mention the inherent cruelty and hiding behind the morality of supposed public good, its defenders put up a financial case. Blood sports, they say, employ rural labour and generate income for rural areas. What they do not mention are the opportunity costs, that is to say the income which might arise from employing the same resources on something less destructive.

Learning the chronology and geography of industries which exploited wildlife for the sake of usable products is also desirable. Specialists in the history of whaling and fur-trapping are among the most sophisticated of environmental historians. In these industries the data go back some centuries and the series are more complete than in most cases. It may be a 'big ask' to expect biologists to go out of their way to acquire historical knowledge but, since it is a key to fluctuations in wildlife numbers, the effort is worth making. The desiderata extend to acquiring a sense of the main phases of growth and depression. An outline economic history is not hard to acquire; it is not, as they say, rocket science.

What is needed is an acceptance of just how far habitats and wildlife numbers, past and present, have been shaped by patterns of land ownership and the demands of the market. Bland statements such as 'man has been a major influence on the natural world' do not go far enough. The relationship benefits from disaggregation. International trade coordinates multiple stories of exploitation. Business enterprises were responsible for great schemes of drainage and land reclamation that had massive repercussions on wildlife. We begin to pick up this saga in sixteenth-century Holland, whence enterprise and engineering skills set out to tame wetland habitats across Europe and beyond. Early seventeenth century Amsterdam made grain, herring, spices and whale oil objects of speculative trading. But

London overtook Amsterdam to become far and away the biggest market for biological goods and had an unprecedented impact on distant ecosystems. We trace some of this through what was offered for sale at Leadenhall market in Victorian times.

To help establish the significance of markets it is useful to consider the environmental implications of staple theory, especially but not exclusively in the lands of new settlement. The main force of the theory (borrowed from development history and economics) is to envisage areas that are cultivated for export products as sequential commodity landscapes. When it is profitable a given crop or an export like timber will stamp its mark on the landscape and a particular array of wild species of animals and plants will become associated with its dominance. A major shift in prices may overturn the association and breed up fresh ecosystems adapted to the landscape formed by the next profitable export. On occasion market conditions will oscillate, meaning that there are historically or economically induced cycles as well as longer-term trends to take into account. To treat past ecosystems as if they were uniform, or as uniformly deteriorating through time, obscures the reality.

This book concentrates heavily though not exclusively on birds. Not only do I know most about them, having been a birdwatcher since my schooldays, but birds are 'portable' in two senses. First, they are mobile, some more than others and at some seasons of the year more than others. Many migrate over global distances. This ensures that there is always something to be seen, which keeps birdwatchers interested all the year round, at least in the northern hemisphere; students of, say, butterflies are less fortunate. Secondly, birds are ubiquitous, which means one can take their study and enjoyment along as part of one's luggage. Birds, then, are the ultimate 'consumables' of the natural world and for an economist that is what matters. More people take an interest in them than in other organisms, especially the creepy-crawlies of the invertebrate kingdom. There is however less interest, dare I say value, in the birds of places one cannot visit. Air travel has drastically cut the number of places the favoured few in the rich countries are unable to reach, and television has made the range of vicarious experiences a global one. But television is a thin substitute for reality, as those tele-naturalists, Chris Packham and David Attenborough, are the first to admit.

At this stage a more concrete summary of the book's contents is in order. By and large I have avoided the fashionable topics of the day, which are too often discussed in generalities. It seems preferable to come as close as possible to what was happening to flora and fauna in the arenas examined. In varying degrees the periods covered run from the sixteenth century to the present. Early chapters try to refute a thesis in the historical literature to the effect that early modern Europe was running into a severe environmental crisis. This does not seem to be true for England or the Low Countries; it appears to result from generalising the experience of marginal northern areas like Denmark and Scotland, and abstracts away from trade as a substitute for a lack of resources in any one place. That neglect is unlikely to persuade the economist or economic historian.

The regions dealt with are mainly Europe (especially England and Holland), North America and East Asia. Because I know it best, I investigate aspects of the environmental history of southern England in greatest detail, while later chapters deal with Europe's expansion, Colonial America and the United States, and East Asia. Full historical data about habitats and wildlife are almost never available for these regions but it is possible to discern differences between trends and cycles in their experience. The observations cited are drawn from as wide a range of sources as possible. In the absence of solid bodies of data, casting the net widely is the best guard against mistaking outlying observations as representing the whole picture. Employing miscellaneous sources also suggests connections that are easily overlooked when relying on narrower and more conventional texts. I do not accept the purist historian's position that everything must be based on primary sources. The scope of the coverage would make that physically impossible. There would be little point in anybody writing if secondary work could never be put to use. Published books and articles contain material that other scholars have sought out and in which they have digested their original researches. Certainly much depends on the questions asked and the models employed. It goes almost without saying that secondary publications have to be read critically, but then so do raw documents.

American material brings to the fore the vexed issue of how unexploited that continent was before white settlement and how reliable are assertions that the indigenous population lived harmlessly alongside nature. East Asia permits an exploration of the way in which biological items can be so much more intensively exploited all around the world when a new region becomes rich — before its traditional uses of natural products have melted away in favour of manufactured goods. Similar impacts occurred during Western economic growth, as can be seen from the contents of nineteenth-century London markets, but the impulse has largely faded in the West. Admittedly, Asian demand for ivory, dietary luxuries and aphrodisiacs is probably not the region's major impact. That is more likely to result from its extensive agricultural and mining schemes in Africa, but little information about their ecological effects has so far filtered into the Western literature.

The book comes up to present times in several places, at greatest length with respect to recent price spikes, the 'farmland grab' in less-developed countries, and the prospects for feeding the world in general. I pay special attention to the stupendous agricultural revolution in Brazil, where soaring grain output has come from areas of dry scrubland rather than the clearances of tropical forest which every newspaper vaunts. As mentioned, the environmental consequences of agricultural intensification on the scale that will be needed to feed the world are not yet clear in any of the big developing regions (chiefly in Latin America and Africa). The prospects do not look comforting, although the gloomier forecasts have a ring of exaggeration. Alarm about 'food security', or rather insecurity, has clearly been overplayed to date, paralleling the familiar lines whereby negative features of environmental change almost inevitably get the billing and positive ones are suppressed. Many problems are less ecological than political — not that this is a guarantee of satisfactory outcomes. Retreat into Trade Protection is a worry and would guarantee unhappy consequences.

At the end I revert to narrower issues of environmental conservation. The final section treats the English case and once again rejects the a- or anti-historical attitude of some conservationists and ecologists. This attitude is exposed where habitats and the associated organisms depend on long-standing, 'traditional' means of managing land and fisheries. I argue that certain of these practices, which are vanishing before our eyes, deserve to be maintained on grounds of cultural history, and indeed would be maintained for that reason in some other countries. No blanket defence of 'tradition' is mounted; witness my objection to the destructiveness of 'traditional' blood sports. But a careful judgement might be made in favour of a few localities where traditional practices preserve specific ecosystems. Unfortunately we live in an era when the legal mind has taken over and compliance with bureaucratic regulations seems to matter more than real functions. I hope that by dwelling on the topic this book may bring about a few reappraisals.

Eric L. Jones
April 2013

ACKNOWLEDGEMENTS

My acknowledgements will be extremely brief when they might be extremely long. Having been a naturalist my entire life, all around the world, I have met and learned from hundreds, maybe thousands, of kind people. Naming them all would be out of the question but I thank them here collectively and gratefully. I will content myself with acknowledging individually a handful of key debts, some of them very old and able to be paid only when the individuals have already left this life. Thus I express my heartfelt gratitude to my geography master, the late Bill Dreghorn, and my naturalist friends the late Colin Tubbs and the late Lew Lewis. Among the living I thank especially John Anderson and Mike Tarrant in Australia and Cliff Irish and Patrick Dillon in England. All of them have been fun to be with in the field and the study or at any rate the pub. John Anderson has also displayed his customary generosity and shrewdness in reading the whole manuscript. So has my wife, Sylvia, who has helped to edit it, as well as coping with the family stresses incurred during the writing of yet another book.

INTRODUCTION

> The land that was once productive of fever and ague, now scarce yields to any in broad England in the weight of its golden harvest. The entomologist is the only person who has cause to lament the change, and he, loyal and patriotic subject as he is, must not repine at even the disappearance of the Large Copper Butterfly, in the face of such vast and magnificent advantages.
>
> Rev. F. O. Morris, on the draining of Whittlesea Mere in 1851

Views of the human impact on nature divide sharply. One school alludes to the benefits of economic growth, the other dwells on cutting nature's carpet into pieces. The economist, Sir John Hicks, refers to the modern world as experiencing a 'secular boom', but the biologist Marston Bates calls it a 'drunken spree'. The world was made by both forces. This book recognises that the environment seems reduced to ill-used fragments yet insists that exploiting natural resources has brought unprecedented gains in well-being: growing population, rising life expectancy and rising standards of living, all of them (amazingly enough) at the same time.

Sections of the conservation community disparage the achievements. Here is the environmental activist, Paul Kingsnorth, on saving the rainforest: 'It was enough to say that it was a home of hundreds of thousands of other species, and that to raze to it to the ground for the sake of toilet paper or soybeans was self-evidently criminal'.[i] Producing toilet paper and soybeans is not a worthy task? To be fair, Kingsnorth does go on to say that destroying old English apple orchards because we want the land for biofuels or housing estates 'is equally, obviously wrong', for less materialistic reasons, because it would be 'a loss to English culture.' Biofuel and houses are unworthy goals too? It would be pleasant to be able to retain wild nature, continue wholesale with quaint, old-fashioned modes of production and still secure a high standard of living — but we cannot achieve everything at once. The three cannot be maximised simultaneously; growth needs to consume resources and a price has to be paid.

The assumption that we can have abundance without environmental cost is commonplace, sometimes expressed directly, more often by implication. It is easy to find pleas for the conservation of nature that lack any sympathy for human welfare. Thus, waxing regretful about the extinction of the black-veined white butterfly, the entomologist, Vere Temple, referred to its former site at Muswell Hill, London, as 'then green country, now a hideous wilderness of brick and stone'.[ii] Housing for the masses is another bad thing? Town and country planning is often crass but the choices that have to be made are not made easier by the 'stop the world' rhetoric of environmentalists.

Naturalists are rarely as open-minded as was F. O. Morris, quoted in the heading to this chapter about producing bread instead of butterflies. Nor are they as even-handed as Atholl Anderson when describing the human occupation of Pacific islands: 'Yet, if this was ecodisaster measured by the loss of island avifaunas, in particular, it was at the same time a kind of "ecotriumph" measured by the success of human colonization'.[iii] Other environmentalists seem resentful about the evidence that welfare is improving while (they assert) ecosystem services are degrading.[iv] The most eminent of biologists may side with the people haters, as when the zoologist, Richard Dawkins, says of the thylacine or Tasmanian tiger, 'maybe they were pests to humans, but humans were bigger pests to them; now there are no thylacines left and a *considerable surplus* of humans... Stop Press: for what may, if verified, be wonderful news, see article by A. M. Douglas...' Consulting the article, we find that it does not describe the reduction in the human population of Tasmania for which Dawkins seems to hanker but a possible rediscovery of the thylacine in West Australia![v]

In the literature of conservation, economics is sometimes dismissed point blank. Where so many individual biologists are concerned, not quite all can be tarred with this brush and one can speak only about tendencies in opinion. Nevertheless, too many make a habit of rejecting economics out of hand, although it is seldom clear which school of economic thought they are targeting or whether they are denying altogether that we live in a world where resources are finite and choices must be made. Every so often, the reader of otherwise interesting books on nature is brought up short by some silly remark. In *Where Have All the Birds Gone?* John Terborgh illustrates the point.[vi] He declares without qualification that economists consider only short-term profits and ignore the long term — he is writing about developing a cattle industry in the Brazilian Amazon — and speaks of 'responsible' accounting that would place a weight on the future equal to that on the present. But it would not be 'responsible' to select a discount rate equating present and future values.

Bjorn Lomborg finds similarly unsatisfactory statements made about the rate of extinction by the biologists Paul Ehrlich and Jared Diamond.[vii] They assert they know for certain we are in an extinction crisis and no research is needed into the actual rate of species loss in order to proclaim this 'fact' to sceptical economists. They are in effect asking society for a blank cheque to prevent a catastrophe whose imminence they trumpet without supporting data. Meanwhile they accuse economists of advocating unlimited human population growth. Maybe there is an economist, or even a handful of them, who do advocate this: the profession is a broad church. But I have never met one nor read of any. Advocacy that may seem to promote this goal will almost always be found not to be doing so at all, though, before Corporate Social Responsibility became fashionable, businessmen could be found making aggressive, profit-is-all, pronouncements.[viii]

Laymen, including biologists, may shudder at the idea of economics and blame the subject for the kaleidoscope of social ills that is always exercising contributors to the media. The prominent biologist, Peter Raven, goes beyond the extreme when he declares that, 'perhaps the most serious single academic problem in the world is the training of economists'.[ix] His complaint seems to be that economists ignore considerations which biologists insist are crucial. Yet the biologists scarcely put their case in testable form. Nor do they distinguish between economists' advice and the actions of politicians. This confusion is what follows when natural science is captured by ideology. The stance of economics, a little logic and a handful of principles should clarify debate.

A dislike of being told that actions have costs seems to lie behind rejections of any need for economic analysis. The prejudice is reversed by those environmentalists who 'know' full well that human actions do have costs and that these costs are too great for the global ecosystem to bear. The fault, they think, is what sub-Marxian terminology would call the 'privileging' of our own species. Costs are taken to be something imposed by humanity, never something imposed on it. Works on biodiversity are prone to assign an infinite value to other forms of life, whatever the burden on people poorer than the writers. To attribute more value to humans than to other organisms is rejected as 'species-ism'. The reader must decide whether he or she really puts the survival of, say, yellow fever and human babies on the same footing. This cannot be shrugged off as a cheap taunt until it is decided where the line should be drawn; no economy can function if each and every population of all species has to be preserved.

Debate in these agonistic terms is fruitless. Arguing with individuals for whom environmentalism is both career ladder and substitute religion gets nowhere since their position is based on self-interest and faith, not science. The real-world balance will have to lie more on the side of people than naturalists would like — more than I like as a lifelong naturalist myself. We simply do not live in a world of infinite resources and what is needed is a calculus of choice that enables decisions to be taken which are sensitive to the relative costs. Since the costs *are* relative there is no point in the gibe that this reduces environmental values to pounds, shillings and pence — of course it does. What alternative has the same comparability? As Julian Simon and Aaron Wildavsky note, policy making is not made easier by environmentalists asserting in one and the same breath that conservation is good for the survival of humanity and demanding limitations on human existence because we are bad for other species. They find these contradictory sentiments expressed by the most influential biologists, such as Jared Diamond.[x] The most thoughtful discussion of economics from a biologist that I have come across is by Steve Nicholls but even he suggests that economists promote totally free markets and ends by applauding restrictions on consumption without explaining how this might work and what the consequences might be for well-being and the distribution of income.[xi]

Besides celebrating greater material well-being and accepting some part of its inevitable downside, we ought to take into account what I have called 'revealed biodiversity'. Economic growth means that more and more people acquire the leisure and wherewithal to 'consume' the natural world. They can afford books, cameras, binoculars, telescopes and travel. The real prices of this equipment have been falling for a long time now. Revealed biodiversity does not refer to the variety or absolute number of species physically present, but to the chance of enjoying the spectacles that do remain, like (say) the increasing number of humpback whales. Opportunities like that count in the eyes of ordinary observers. Bishop Berkeley's famous tree in the quad may as well not be there for people who cannot travel to see it. Experiencing the splendours of the outdoors through television is no substitute for tiny experiences of one's own. In England, television means that a generation of children is withdrawing, or being withdrawn, from the countryside.[xii]

Boundless opportunities still exist. Wealth makes the mass enjoyment of nature possible, besides endowing the conservation movement with nature reserves and employing unheard-of numbers of people as wardens, professional naturalists and (let it be said) conservation bureaucrats. This point may be missed by writers on the topic, who have had great opportunities themselves and are concerned about their continued access to remote places and extreme rarities: the cloud forest syndrome. Natural history was once the hobby of people seeking solitude — it was part of my own inspiration. We may dislike having to take our place in lines of telescope-wielding twitchers but exclusiveness is neither a generous attitude of mind nor any longer a feasible one.

The suggestion that humanity is better off as a result of its affluence and the new technologies of communication and transportation will scarcely weigh with anyone disposed to think that all species are of equal value. They will fear that little will be left for the newly prosperous to watch. What then will more leisure and greater mobility be worth? Will all that remains be flies, sparrows and rats, creeping around the weedy bombsites of human occupation? But even if such fears were justified they would be far from the attitudes of dispassionate scientists who find as much of interest in backyard weeds as in the orchids of the cloud forest. In any case, the approach places too much emphasis on the concreteness of species, the definition of which is uncertain at the margin.

A contrary point has been vigorously put: that the mass extinction of species does not really matter. Calculations by Robert May, once the U.K. government's chief scientist, and another Oxford ecologist, Sean Nee, show that 80% of the underlying tree of life would persist even with a 95% species loss. About 1,500 species of birds may perhaps become extinct within a few decades but the group as a whole will flourish because birds are adaptable and thrive in man-made habitats. Their biomass, if not their variety, will continue to do well. And the fearful estimates of species loss have been challenged as too high.

The United States Supreme Court found in 1978 that the intention of the Endangered Species Act (1973) was to halt extinctions at any cost. But attempting to preserve every species is impossibly expensive and cannot confidently be left in the hands of government agencies. The issue caught public attention because of the cost of the effort at saving the snail darter (a fish from Tennessee). Not every endangered species could receive such lavish treatment. A Senate 'God Committee' had to be formed to pick and choose among them — and meanwhile the snail darter was found in more places than anticipated. The goal of preventing the extinction of any organism at all is unrealistic: the effort (which failed) to save the dusky seaside sparrow cost virtually the whole of the $1.3 million appropriated for acquiring reserves. Furthermore it could be argued that we seldom know enough about habitat requirements to judge exactly where money should be spent. Ecology is a relatively young science and has tended to take things as it finds them, without paying much attention to what the past may tell us about fluctuations in the natural world, even when the data do run far enough back in time to separate cycle from trend.

PROCESSES

An episode where desperately needed historical data were lacking was the alarm in the late twentieth century about the damage wrought to coral reefs by population explosions of the crown-of-thorns starfish. This struck at the Great Barrier Reef, which is important to Australia's tourist industry. The ecological interactions are complicated but it turns out the species has erupted periodically on the Great Barrier Reef, the suggestion being that the cycle occurs once every 400 years.[xiii] About the same time, there was hand-wringing about the die-back of the eel-grass, *Zostera*, in Westernport, a large tidal bay near Melbourne.[xiv] Eel-grass is an important shelter for fish and crustaceans, and is grazed by birds. This die-back, too, seems not to have been a once-and-for-all plunge into the abyss of extinction but to be cyclical, as it also seems to have been in Chesapeake Bay in the United States and the Solent in England.

The situation in the Solent with respect to Zostera was investigated by that indefatigable ecological historian, Colin Tubbs.[xv] He found that the plant had been abundant before about 1930 and traced observations as far back as that great romantic landscape writer of the late eighteenth century, Rev. William Gilpin, whose parish at Boldre in the New Forest lay close to the Solent. But about 1930, a 'wasting disease' hit Zostera beds on both sides of the Atlantic. Fishermen, naturalists and observers who simply resented change in the environment were upset by this and assumed they were seeing an entirely novel threat of extinction for the plant. Yet after the 1950s recovery began and since then has been considerable in some parts of the Solent.

A little ecological history thus helps to temper knee-jerk responses to environmental change. It controls assumptions that the environment was always 'there' in a fixed form until disrupted by the hand of modern man. That unhistorical attitude sanctions attempts at preserving, recovering or recreating species that may have flourished only in the recent past. A piece of woodland at East Blean, Kent, is said 'always' to have been coppiced for pulp wood for a paper mill in Sittingbourne.[xvi] When the mill closed during the recession of the 1980s, coppicing ceased. The sunny glades hitherto filled with cow-wheat were swamped by tree growth, closing out the colony of a butterfly, the heath fritillary, which depended on the cow-wheat. The loss of butterflies that are intolerant of shade is a familiar outcome when coppicing ceases.

The heath fritillary has become one of the country's rarest butterflies. Unfortunate though its loss is at East Blean, it is unlikely 'always' to have been present in numbers. Indeed 'always' can be guaranteed to be wrong. Historically speaking, coppicing was a fairly recent practice, though it became widespread and is of great interest in its own right. The website for the history of the Sittingbourne mill says that the first paper-maker was mentioned only in 1708. If paper-making was responsible for creating the habitat for heath fritillaries, what was their status before then? Perhaps the glades of the previous wood-pasture environment had suited them and coppice was a lucky substitute. There is no fixed, unchanging fauna and flora whose loss we should lament. From the conservation standpoint, as large a variety of habitats and habitat-forming processes as possible is desirable. This is a form of insurance when our predictions are so uncertain. In addition, all processes of change offer experiments or opportunities for study; playing favourites among species ought to be beside the point.

The contingent element in the numbers and distribution of organisms of all types in habitats influenced by man should be kept in mind. Apart from fluctuations arising from natural causes, such as the weather, and the almost unpredictable knock-on effects of inter-specific competition, habitats tend to be unstable. Even in the largest cities, human action creates fresh habitats, and it has been suggested that the additions made to fauna and flora in the course of transforming the American continent represent a net gain.[xvii] In the oceans, it can be seen that waste from the North Atlantic fisheries was responsible for the spread and growing numbers of fulmars, the history of which was meticulously documented by James Fisher. In a couple of centuries, fulmars prospected and colonised over 2,300 miles, right from Grimsey, Iceland, to Sussex. The offal from whaling ships that entered northern waters in the seventeenth century started the spread, and when the right whale was fished out, waste from trawlers took over. The fulmar is not alone as an anthropogenic species — in the plant world I would instance gorse.[xviii] Plenty of examples are matters of record and doubtless others await their historians.

The negative effects of hunting are more prominently reported. Reports have a habit of being skewed in favour of the losses. The extinction of the great auk is infinitely better known than the increase of the fulmar, despite the fact that the latter was 'the most spectacular and continuous spread of any wild bird known — apart from the colonisations of the starling and sparrow that were started by human introduction.'[xix] The rush to judgement that the human influence is necessarily malign is unwise. In longer perspective, it is better thought of as a source of disequilibria, a speeding up of the processes of disturbance. Although observers are unforgiving about what they perceive as upsets to familiar arrangements during their own lifetimes, habitats and distributions tend to settle down over the years. In his unfashionably positive view of humanity's future, Rene Dubos refers to the wonderful harmony in many parts of Europe as a result of what has been called the 'wooing of the earth'.[xx] Dubos grew up in the Ile de France; I grew up in Hampshire and agree with him about the disciplined charm that can be reflected in humanised landscapes. Like the architectural harmony of townscapes, it can be an unintended consequence of juxtaposing diverse elements.

More general points include how enormous the human impact has been, how impossible to reverse in most cases, and yet how liveable are the sometimes depauperate ecosystems that have resulted. People who are not especially interested in nature show a considerable tolerance for environments less rich in natural variety than those their forebears knew. Humanised landscapes have proved quite stable and contain the added interest and aesthetic pleasures of parks, gardens, buildings and other artefacts; with these many people are content. Nature was long regarded as unappealing and hostile by those who had to live cheek by jowl with it, Gilpin's enthusiasm for wild scenery at the end of the eighteenth century being the first sign of altered opinion. The change came first among the well-to-do who could afford to travel to the Lake District or the Scottish Highlands. The opportunities are much wider now.

Enthusiasts for nature are more numerous today than ever, but a majority of people in the developed countries, school teachers among them, are no longer able to identify common birds or flowers. Modern populations have unprecedented chances to enjoy nature but those without a special taste for it are detached from the countryside as never before. Related to this, the use of maps is retreating in favour of SatNav systems in cars, a form of self-imposed de-skilling especially lamentable in the United Kingdom, where the Ordnance Survey produces such elegant maps crammed with historical and other information. University lecturers in the field sciences lament how few freshmen possess the taxonomic skills that were once the gateway to subjects like botany. An Oxford geologist told me that undergraduates positively loathe fieldwork, preferring computer modelling. From the strictly scientific point of view this may be acceptable, though the lack of curiosity bodes ill. If these are the traits of university people, at least it becomes understandable that the mass of the population is content with what is called the 'Crete Scenario', meaning heavily used landscapes with limited natural variety, provided only that they sense wild places still exist 'out there' and their children may see exotic creatures in the zoo.

Whether traditional field work in ecology will long persist is questionable. The United States Congress has funded NEON (National Ecological Observatory Network) at a cost equivalent to that of a space probe.[xxi] It bids to turn ecology into Big Science, like other sciences that now generate vast amounts of data to be processed by computer. A total of 15,000 sensors placed in each of 20 domains, together with aerial surveys, will collect 500 types of data, amounting to 200 terabytes per annum. This is four times that which was collected in the first twenty years of operating the Hubble telescope. The days of field glasses and butterfly nets may be numbered, says the *Economist*. Students of ecology are likely to turn from field work to crunching data. Whether this will result in more original findings per dollar expended than traditional research like that at Wytham Woods ('Oxford's Outdoor Laboratory') remains to be seen, as does the impact on recruiting the idiosyncratic characters and original minds of the students that ecology has hitherto attracted.[xxii] Either way, the sub-discipline of terrestrial ecological history is likely to languish as a poor relation, characterised by a few intensive studies and vague 'global' essays amidst a mass of fairly haphazard findings.

ECOLOGICAL REARRANGEMENTS

An original intention for this book was to set against one another historical instances of loss and gain in the natural world. But after compiling long, long lists of examples on both fronts, this appeared futile. What could the baseline be and how might the relative costs and benefits be compared in more than an indicative fashion, given the absence of a metric like market price — and for that data are largely missing. The world is already a 'witless menagerie' in which species from here are inextricably mingled with ones from there. Conceivably, habitats in the most remote places may be preserved intact, though 'intact' is historically doubtful term given that much even of the Amazonian rainforest grows on land once cleared for farming. Most of the earth is occupied by syncretic ecosystems resulting from innumerable human interventions, deliberate and unintended.

The classic paper is by Peter Vitousek on the 'Human Domination of Earth's Ecosystems'. He indicates that domination is almost complete.[xxiii] Beware, however, his technically correct conclusion that about one-quarter of all bird species have been driven to extinction; it may give a misleading impression. Modern readers have been conditioned to expect that such a statement must refer to Europe's expansion overseas, involving massive agricultural reclamation in the less-developed countries. But the lion's share of the losses occurred in prehistory, as the result of pre-European colonisation in the Pacific islands: in the process 20% of the worldwide avifauna was lost, the majority being flightless rails confined to specific islands.[xxiv] Human action definitely — but not easily assimilated to a narrative dwelling on the supposed ills of European colonialism. Island ecosystems were particularly vulnerable to fire-stick assaults on their vegetation and the introduction of predator species. Two wrongs do not make a right, yet the remaining 5% of the stock of birds rendered extinct in the great land-masses seems small by comparison, given the extent, duration and ferocity of colonial occupation. A reason for this has been the staggered nature of forest clearances, leaving reservoirs of different species able to recolonize affected areas.[xxv]

The land surface of much of the earth has been remade, with all this implies for plant and animal communities. Plain as a pikestaff in urban areas, the effect is more subtle in the countryside, where present arrangements are readily taken at face value. When the cycle of fertility transfers is considered, the continual remodelling effect of urban markets, for example, makes itself clear. Thames barges brought straw and feedstuffs for London's large horse population, besides food for humans. Dung from the horses was spread on market gardens in Middlesex, the crops grown were consumed by London's inhabitants, and their excreta were voided down the Thames.[xxvi] In the long term, the effect was to extract fertility from the supplying districts, such as Essex, although they eventually received some offset from an unlikely source, the import of guano (seabird droppings) from Peru.

The wheels of fertility transfer were complicated, overlapping, and ultimately global.

Fertility cycles may be found everywhere. English parishes are usually most fertile in the gardens around the church, where manure was continually brought in, and in nearby fields where for many centuries sheep were folded at night after being grazed on more distant grassland. Outer parts of a parish lost fertility and it is no accident that it is on their fringes are the farms with names like Starveall. The intense significance of fertility for crop growing is also shown by the typical clause in farm leases forbidding tenants from ploughing up old pasture. Similarly, the brickmaking leases common in Middlesex required the occupiers to remove the topsoil and replace it after completing their work.[xxvii] In recent times, the opposite concern manifests itself: when artificial fertilisers are (too) freely applied on farmland, waterways can become eutrophic, that is to say covered with a scum of algae that live on the excess nutrients. Interactions were and are endless.

What might be called deeper history provides a record of clearance, massive deforestation, land drainage and transfers around the world of innumerable species of animal and plant. The process has recurred over millennia, slow or fast, and included the great translocations like those that created the neo-Europes, to whose reformulated biology Alfred Crosby drew such fruitful attention.[xxviii] Overawing though the overseas expansion of the European peoples and their 'living entourages' was, it was only a particularly vigorous episode in the series of overlapping colonisations by human societies as far back as one cares to look. Common practice was to start by trying to acclimatise species from the colonists' homeland and to end by introducing or creating the conditions for fresh organisms in numberless variety.

All over the world, there has been trading in animal parts for ornamentation and other uses, and the collecting of eggs for food.[xxix] The unbridled destruction of egrets, terns and grebes for plumes and other adornments of women's hats, to suit the Western fashion industry, spurred early moves to organise bird protection. Plumes were sinuous and sexy and the market for them was so big it seems almost unsurprising that consignments of plumes were the most valuable part of the cargo that went down with the *Titanic* in 1912. In a familiar sequence from the 'lucky dip' gathering of wild resources to growing them deliberately, feathers acquired by hunting were supplemented by ostrich feathers farmed in South Africa, whence they were presumably difficult to fetch during the First World War. The 'feather crash' of late winter, 1914, killed the trade.[xxx] There was also a change in fashion away from ornamented hats, though the precise chronology and which effect predominated is unclear; they probably reinforced one another. In Britain, the Plume Act of 1921 helped to suppress any resumption.

The more compelling effect of colonisation was to establish wide monocultures of crops, at least in the European diaspora, once it became possible to ship produce back to the 'metropolis' on a grand scale. Caught up in the transfer were species so well adapted to the monocultures that they became pests. Opportunistic local species joined in. Meanwhile native game animals (not to mention natives) were slaughtered to the brink of extinction or beyond, after which they were replaced by introductions. Livestock farming required introducing and breeding domestic animals in enormous numbers; by 1880 the hog population in the Middle West alone had reached 50 million.[xxxi] The overall implications of these gigantic changes for resident species and the initial vegetative cover are scarcely calculable.

Natural resources have often been depicted as common property resources where the 'tragedy of the commons' unfolds — the theory is that there seems no reason for restraining oneself when others do not. This traduces the institutional creativity of our rustic ancestors. On farmland in long-settled countries there was seldom any 'tragedy'; villagers successfully restricted each other's use of common property. Their needs and energies burst out in other directions — they could act greedily towards wild species in which no one had specific rights. International merchants were prompted to behave likewise by the market expansion of Victorian times. In addition to direct assaults on the wild, developed countries reshaped their own ecologies by housing large numbers of pets. The growth of pet numbers was and is partly a function of rising real incomes, or rather the diminishing marginal utility of real income, as may be seen from the extravagant mollycoddling of pet dogs in China, where there was no tradition of keeping such a non-utilitarian animal. When pets are free to roam, as the enormous cat population is, the consequences for wildlife are tragic. Economic change, new technology, consumption habits and shifting fashion create astounding stews of ecological ramifications.

Dubos cites Carl Sauer as observing that it is recent settlements which are worn out, not the lands of old civilisations, at least outside the semi-arid and arid zones. The regular rainfall of Western Europe, Japan and some other parts of Asia lets their soils recover easily and already by the mid-1970s Dubos was able to list other forms of recovery — of polluted lakes and polluted air — brought about by environmental policies in a number of Western countries. Whether or not future challenges will also be resolved is an open question but historical experience in the West suggests they can be met, since in rich societies conservation is a consumption good.

History is full of instances of the translocation of species through many sorts of mechanism. Usually the press headlines losses in apocalyptic terms while recoveries appear only as snippets. Notwithstanding the cutting of wild land into smaller and smaller pieces and the decline of more species than can manage to recover, the situation is not quite as bleak as portrayed by alarmists blindly devoted to maximum biodiversity. Their view does not give enough credit to the growing incomes that support 'revealed biodiversity', nor does it admit the liveability of humanised landscapes once people and nature have reached new equilibria.

Such equilibria are of course temporary. Everything depends on how long is the period of observation. As Piet Nienhuis points out, 'the term "native" is used for organisms which have been native to a particular geographical area in historical times. However, the concept of "historical times" is rarely defined.'[xxxii] It cannot be reiterated too often that changes are always poking their way in. A debate has erupted over the role and designation of invasive species, a category that has already given rise to the sub-discipline of invasion biology. 'Alien' species are accused of ousting natives and greatly reducing biodiversity — however, newcomers are usually omitted from the species count until they become naturalised in turn. The language used is emotive and designed to justify the often unavailing expense of campaigns to eradicate incomers. The most creative response has been Jackson Landers' proposal to hunt and eat even the more exotic among them.[xxxiii] There are cases where incomers have expanded fast, distorted existing ecosystems and imposed serious costs on forestry or water supplies. But so have eruptions of native species. Any rapid spread can be looked on as a disruptive technology that disturbs some prior equilibrium.

The baseline in such cases is seldom specified and justified, although it is openly stated in Australia, where the purity of the vegetation has become a fetish. The baseline there is set just before the contact period, the European arrival. The aim is a botanical ethnic cleansing meant to restore the vegetation of just before white settlement, as if aboriginal populations had not previously modified their environments. This chauvinistic policy leads to expensive efforts at plant extermination or restoration. In the United States, where in a fit of irony some state emblems are non-native flowers, policies like this have been deprecated, according to

Neyfakh, as 'ecological nationalism' and comparison has been made with the anti-immigration movement.[xxxiv] Neyfakh goes so far as to assert that those who oppose invasion ecology are usually on the political right and oppose spending on conservation. This is as unproven as the claim that sceptics about global warming are in the pay of the oil companies! For the purposes of land management, each case needs to be considered on its merits and the agonistic tones of some literature avoided. A more persuasive contribution to the debate renders the wrangling irrelevant by urging that the distinction between native and non-native species is artificial, as well as unhelpful in a setting of syncretic and ever-altering ecosystems.[xxxv]

Old-settled lands have had thousands of years to absorb interlopers, which tend to be much more disruptive in the less disturbed lands of recent settlement.[xxxvi] Part of the reason is that invasive species find it easier to penetrate disturbed land, which is more a feature of areas of recent colonisation than of long-settled ones. Nienhuis notes that the inherent susceptibility of disturbed areas to invasions 'has become an accepted cultural-historic phenomenon.' The dimension of culture history is little explored in typical work on ecology.

Better targeted efforts at conservation could provide ample opportunities for enjoying nature in the most densely settled countries, like England. Living as I do in part of that realm (the Cotswolds) heavily damaged by subsidised agriculture and from which the public is largely excluded by virtue of game preservation, I do not deny that environmental impoverishment may not persist or even worsen. It is a matter of political will. The remedy is curbing excesses and devoting a larger but not insupportable share of the gains of economic growth to the restoration of nature.

ENDNOTES

[i] P. Kingsnorth, *Real England* (London: Portobello, 2009), p. 258.

[ii] V. Temple, *Butterflies and Moths in Britain* (London: Batsford, 1945), p. 13.

[iii] A. Anderson, 'Faunal collapse, landscape change, and settlement history in Remote Oceania,' *World Archaeology*, **33** (3) (2002), p. 375.

[iv] C. Raudsepp-Hearne *et al.*, 'Untangling the environmentalist's paradox: Why is human well-being increasing as ecosystem services degrade?' *BioScience*, **60** (2010), pp. 567–589.

[v] R. Dawkins, *The Blind Watchmaker* (Harlow, Essex: Longman Scientific and Technical, 1986), p. 105, italics supplied; A. M. Douglas, *New Scientist*, 24 April 1986.

[vi] J. Terborgh, *Where Have All the Birds Gone?* (Princeton: Princeton University Press, 1989), p. 181.

[vii] B. Lomborg, *The Skeptical Environmentalist* (Cambridge: Cambridge University Press, 2001), p. 256.

[viii] See, e.g., in S. Nicholls, *Paradise Found: Nature in America at the Time of Discovery* (Chicago: University of Chicago Press, 2009), p. 170.

[ix] Quoted in J. L. Simon (ed.), *The State of Humanity* (Oxford: Basil Blackwell, 1995), p. 357.

[x] *Ibid.*

[xi] S. Nicholls, *Paradise Found*, especially, pp. 451–458.

[xii] It is said that an Oxford dictionary for children has dropped words like 'acorn' as irrelevant to the present generation. I cannot bring myself to check, for fear it may be true.

[xiii] L. Kaufman and K. Mallory (eds.), *The Last Extinction* (Cambridge, Massachusetts: The MIT Press, 1986), p. 21.

[xiv] Despite its location east of Melbourne, Westernport was the westernmost point reached by the first explorers from Sydney.

[xv] C. Tubbs, *The Ecology, Conservation and History of the Solent* (Chichester: Packard Publishing, 1999), pp. 116–126.

[xvi] P. Barkham, *The Butterfly Isles* (London: Granta, 2010), p. 197.

[xvii] R. S. R. Fitter, *London's Natural History* (London: Collins New Naturalist, 1945); B. Gilbert, *Our Nature* (Lincoln, Nebraska: University of Nebraska Press, 1986), p. 42.

[xviii] For gorse see E. L. Jones and C. R. Tubbs, 'Vegetation of sites of previous cultivation in the new forest,' *Nature*, **1988** (1963), pp. 977–978.

[xix] J. Fisher, *The Shell Bird Book* (London: Ebury Press and Michael Joseph, 1966), p. 116.

[xx] R. Dubos, *A God Within: A Positive View of Mankind's Future* (London: Abacus, 1976), pp. 100ff.

[xxi] *The Economist*, 25 August 2012.

[xxii] See P. S. Savill *et al.*, *Wytham Woods: Oxford's Ecological Laboratory* (Oxford: Oxford University Press, 2010).

[xxiii] P. M. Vitousek, 'Human domination of earth's ecosystems,' *Science*, **277** (1997), pp. 494–499.

[xxiv] D. W. Steadman, 'Prehistoric extinctions of Pacific birds: Biodiversity meets zooarchaeology,' *Science*, **267** (1995), pp. 1123–1131.

[xxv] S. Nicholls, *Paradise Found*, p. 150.

[xxvi] M. Thick, *The Neat House Gardens: Early Market Gardening Around London* (Totnes: Prospect Books, 1998), pp. 101–103.

[xxvii] P. Hounsell, 'Brickmaking in the Hayes area in the 19th century,' *Hayes and Hartington Local History Society Journal*, No. 84 (Autumn 2011), p. 11.

[xxviii] A. Crosby, *The Columbian Exchange: Biological and Cultural Consequences of 1492* (Westport, Conn.: Greenwood Press, 1972).
[xxix] E. L. Jones, 'Utilisation by Man,' in B. Campbell and E. Lack (eds.), *A Dictionary of Birds* (Vermillion, S. D.: Buteo Books, 1985), pp. 617–618.
[xxx] S. A. Stein, *Plumes: Ostrich Feathers, Jews, and a Lost World of Global Commerce* (New Haven: Yale University Press, 2010).
[xxxi] L. Watson, *The Whole Hog* (London: Profile, 2005), p. 172.
[xxxii] P. H. Nienhuis, *Environmental History of the Rhine-Meuse Delta* (New York: Springer, 2008), pp. 452–453.
[xxxiii] J. Landers, *Eating Aliens: One Man's Adventures Hunting Invasive Animal Species* (North Adams, Mass.: Storey Publishing, 2012).
[xxxiv] L. Neyfakh, 'The invasive species war,' www.boston.com, 2 August 2011.
[xxxv] M. Davis *et al.*, 'Don't judge species on their origin,' *Nature*, **478** (2011), pp. 151–154.
[xxxvi] P. H. Nienhuis, *Rhine-Meuse Delta*, pp. 452ff.

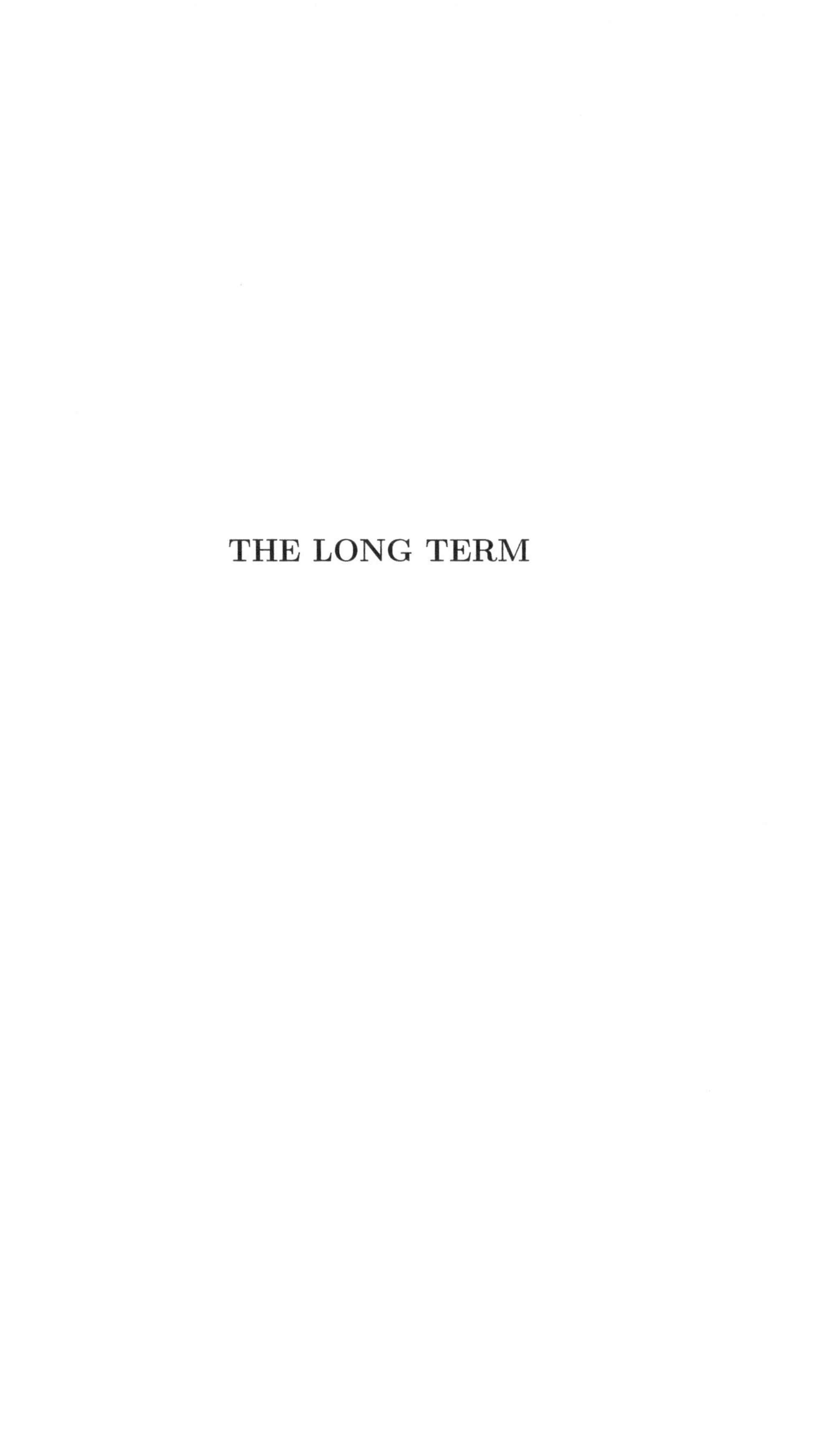

THE LONG TERM

1. CRISIS

> I think biology furnishes the necessary background against which social studies must be built. And we must take care that these various fields of inquiry do not really develop as separate worlds.
>
> Marston Bates, *Where Winter Never Comes* (1963)

One of Adam Smith's most celebrated remarks was that the establishment of the European colonies in America and the West Indies arose from 'no necessity'. Exploration was not driven by ecological crisis at home. Certainly, efforts to expand the resource base were proceeding overland, in Eastern Europe and across the Russian steppes. But Europe did not strictly need a fresh trove of real resources like food, fish and timber, welcome though they might be, nor were the first explorers seeking them. There was still scope in the landmass of which Europe was part.

The explorers who set sail to Asia and the Americas claimed to be motivated by spiritual concerns and were obviously impelled by an interest in luxury items and a lust for treasure. When Vasco da Gama set foot in India, he announced he wished to find Christians and spices. His supposed quest for lost co-religionists like Prester John was, however, little more than geopolitical dreaming. If Christians were found, no one was going to repatriate them; the most that could be hoped was their help in encircling Islam. Aside from this fanciful aim, the explorers' minds were fogged by the mercantilist mirage that gold and silver were wealth and wealth was gold and silver — and, as Smith said, 'the like chimerical views'. Any commodity that could be fetched home to Europe would have to be of high value but little bulk, since nothing more was manageable in the tiny vessels of the fifteenth century. Distant supplies of timber or grain were beyond the shipping capacity of the time.

When Europeans and Americans did eventually find a market in China for the products of the southern oceans, what they took there were the pelts of fur seals. They fell back on what was available, what they could carry and what they could sell. They were not driven by prospects of establishing great industries or founding agricultural enterprises on the resources of the outer world. They came closest to such a prospect in the Atlantic trade, but that was long after Columbus. Developing the Americas was neither quick nor easy. In the eighteenth century, colonial Americans sold grain and packed pork to markets in Iberia, yet only in the late nineteenth century did it become feasible to ship utilitarian goods back to Europe on a scale that transformed whole economies.

Nevertheless, a school of thought exists, one that brooks no dissent, which holds that Europe's expansion really was necessary, and more than necessary, essential. Scarcity was certain to bite in Europe, it asserts; the problem was merely being deferred while the continent outgrew and degraded its own resource base. On this view there may have been 'no necessity' for overseas ventures during the demographic slack of the fifteenth century, but an expanding Europe would soon come to require — so it is claimed — a flood of distant resources neither anticipated nor capable of being transported in Columbus's day. These resources amounted to energy and nutriment.

Traditional raw materials were mostly organic — things that could be grown. Europe found them inelastic in supply, unable to sustain indefinite population growth, much less produce a rise in the standard of living. The eventual response was indeed the massive import of resources and the substitution of inorganic for organic sources of heat. This meant mining coal and finally using it to manufacture inorganic chemicals. These epic shifts were supplemented by importing food and raw materials from outside the continent. Without innovations at home and exploitation abroad, runs the argument, a crisis of food and energy would have overwhelmed Europe by 1800 or 1820, and the early signs were visible much earlier. If this story could be relied on, we would observe massive effects on habitats, fauna and flora, not least because desperate people would have scoured the countryside for anything to eat or put on the fire. Very poor people did forage hard, especially in bad winters, just as the impoverished are very hard on the environment in less developed countries today. But claims of an all-encompassing early modern European crisis seem to reflect a reading back into the past of modern fears about global resource scarcity.

The case that an environmental crisis was under way throughout Europe is made most forcefully by Thorkild Kjaergaard, who generalises the example of Denmark.[i] We need to enter a little into the details in order to gauge how real and serious the problem may have been. It is, therefore, unavoidable to discuss early modern and early industrial history in ways that will be unfamiliar to people trained in biology. Establishing what the trends and fluctuations were is vital in order to prevent misunderstandings, in particular to show how easy it is to select an inappropriate baseline for ecological history and misinterpret later developments.

Kjaergaard starts with the effects of the great depopulation caused by the Black Death in the mid-fourteenth century, which reduced the demand for farmland and meant that, by the end of the next century, Denmark had plenty of second-growth timber. Since the population was then less dense than it had been during the High Middle Ages, it may have been able to devote more land to livestock. Consuming additional meat and being able to deploy more draught animals would have improved living standards.

These changes, though somewhat conjectural, are plausible enough. Yet the aftershocks of the plague eventually wore out and, after 1500, the population was growing again. The increasing number of people was not easy to feed with the restricted agricultural technology of the day. More people seeking more farmland meant woodland clearance. Kjaergaard also thinks that timber was burned up, so to speak, by building lighthouses, palaces and a navy, although it is hard to believe these were major causes of deforestation. The most creative response was the substitution of stone for wood as a building material — and keeping stone buildings warm took more fuel. Clearing the forests exposed the soil to leaching and loss of nutrients.

Nitrogen was scarce everywhere and was needed on the cropland, meaning that it had to be transferred to the arable from the pasture, which as a result could barely support the large but half-starved national herd of cattle. Moreover, the shock of cattle plague in the 1740s halved the size of the herd. Meanwhile, a lack of field drainage limited the length of time during which farmers could work the land, thus curtailing the season for growing crops. The result was a shortage of fuel, manure and food. Yet, given Kjaergaard's thesis, it is paradoxical that, even though Denmark's population grew during the seventeenth century, agriculture was not an attractive investment. The courtiers did not buy land, reported Robert Molesworth in 1692, but placed their money in the banks of Amsterdam and Hamburg.[ii]

Because deforestation had made fuel scarce, cowpats were dried and burned instead of being used for manure, exacerbating the common European problem of declining soil fertility. This opinion is not Kjaergaard's alone. 'The agricultural history of Western Europe', says Smout, picturesquely but not unreasonably, 'can be written in terms of efforts to plug nature's leaking cornucopia with sufficient manure'.[iii] In accounting for this sorry state of affairs, Kjaergaard extends his argument to the effects of early modern state-building and the resultant tax burden. During the sixteenth and seventeenth centuries, Denmark created a militaristic and high-tax state. From the 1660s to the 1740s, taxes were so high that farmers were obliged to increase output and sell more produce to meet their dues. The overall result is portrayed as a curious conjunction of ecologically damaging over-production in agriculture, product prices that did not fall despite the greater output and an absence of technological innovation. Lack of innovation accompanied by a growing population led to crisis. Existing organic methods could not cope: by the early eighteenth century, there was an energy crisis not only in Denmark but supposedly throughout Europe.

Is Denmark really a satisfactory model for the remainder of Europe? Its experience does not look much like that of the United Provinces or Lowland England. Features of its circumstances do not really add up. First, Danish agriculture ought to have been profitable but seems not to have attracted investment. Second, Denmark was only a small economy, with a population of 700,000 in 1600 (England and Wales: 4,250,000) and one million in 1800 (England and Wales: 9,250,000). It lay well to the north in Europe, with all this implied for climate and plant growth. Similarly, northern areas of England were fairly marginal and their economies did not grow rapidly until the eighteenth and nineteenth centuries. During the seventeenth century, Smout tells us, the assessed wealth of England north of the Humber, one-fifth of the whole country's land surface, sometimes barely equalled that of a single southern county, Wiltshire.[iv] He wisely adds that this estimate derives from tax records and in the north taxes may have been systematically avoided, but nevertheless he accepts that there must have been a large disparity between north and south.

It is hard to escape the fact that areas like Denmark or the north of England were peripheral to the agriculturally more benign and economically developing parts of Europe. Marginal areas were marginal. If they failed to introduce striking innovations — the desirability of innovation never guarantees a response or the history of the world would have been very different — they were likely to suffer stress when their populations were growing. Thus their environmental histories are likely to bias downwards our understanding of the pre-modern situation.

In any event, 'crisis' is scarcely an apt label for a phase which by Kjaergaard's reckoning lasted for a century until 1800 to 1820, by which time he claims Europe was right on the brink of being overwhelmed by ecological troubles. I suspect some of his impression mistakes distributional changes, the poverty reflecting who-got-what at particular junctures, for overall prospects. On the contrary, the limitations of the organic economy were overcome. Kjaergaard's opinion is that the means of escape was inorganic — the adoption of coal and the greatly increased quantities of iron that could be made by using it, coupled with raising agricultural productivity by sowing clover. Clover is a legume that fixes its own nitrogen (and is not, of course, inorganic). It had been sown in Europe as early as A.D. 1000, but there was an immense delay before its widespread use in the seventeenth and eighteenth centuries, when a nitrogen deficit was recognised. Thereafter clover became indispensable for producing food. 'In the nineteenth century clover was the agricultural equivalent of coal', Kjaergaard states.[v] Coal and clover transformed the land: 'the romantic nineteenth century landscape was a clover landscape', with all this entailed for a buzz of insect life rarely heard in modern fields.

Also waiting in the wings was printing, which facilitated the diffusion of knowledge about new techniques. As a result, the standard of living rose throughout Western Europe. Crisis was averted. Not to be beaten, Kjaergaard promises us that this was, and is, a false dawn (although we may note that the dawn has already lasted longer than the crisis). He is convinced that the new technologies were, and are, interim measures, and by the end of the twenty-first century our economic and environmental system must collapse. This forecast is untestable (as yet) and unpersuasive, because it denies the possibility of discovering fresh resources or inventing fresh technologies. It is as if, despite all previous experience and the research and ingenuity of modern science, history must now come to a halt. Why now, in the present century? This is the fallacy of believing that the course of history leads up to one's own day and there grinds to a halt.

Modern industrialists take no such closed position and anticipate continual advances. Admittedly, they are said, by one Nobel Prize winner, to go out of their way to steal the inventions of others by writing multiple similar patents and, if the original inventor sues, threatening to break him through unsustainable legal costs.[vi] Publishers' contracts sometimes insist on acquiring copyright over forms of technology yet to be discovered! However, none of this would seem to envisage scientific progress coming to a stop. Given that the escape from earlier resource traps depended on unforeseen advances, such as the production of synthetic nitrogen, an assumption that technological change has ceased would appear inconsistent.[vii] It is also a recipe for inaction.

Kjaergaard makes everything in the early modern organic regime fit distressingly together. The future was dystopian, as it is in 90% of environmentalists' forecasts. The organic economy is said to have been unable to cope: it could neither have supported nineteenth-century population growth nor delivered the rising living standards attained at that time. A step-change was essential for warding off disaster. Coal and clover were the novelties said to have turned the trick. Waving the magic wand of coal-and-clover delivered Europe from the dead end to which history purportedly was and, in Kjaergaard's view, still is tending. For all the neatness of the story, the suspicion must linger that the Danish predicament is being over-extended and the early modern world painted in unjustifiably sombre colours.

Our impressions of the past are dominated by the Victorian achievement, an era that draws a thick veil across earlier times and makes them hard to discern. Narrators of industrialisation tend to understate the vigour of the previous economy, at least in parts of North-West Europe, and paint too glib a picture of resource scarcity. The result is an emphasis, amounting to a fixation, on iron, coal and industry in the coal districts, as well as on the massive import of cereals from outside Europe. This is not to suggest, of course, that the adoption of coal, cheaper iron and the use of clover were anything less than game-changers, though an exclusive emphasis on clover short-changes and misdates the range of agricultural innovations being introduced from the seventeenth century.

These factors were key components of the emerging industrial world. This is in no doubt: the error is that their scale and novelty are permitted to overshadow earlier times. Their adoption *en masse* is dated too early and made to seem too abrupt, with the effect of minimising the subtler changes of the lead-in period. Knowing that inorganic resources and new technologies were indispensable to the nineteenth century's combination of rising population, city growth and better living standards has an unfortunate effect. It leads to portraying the methods of earlier times as ultra-weak and incapable of development. That they were unlikely to have been able to support full nineteenth-century growth is very different from thinking the pre-industrial European economy was either static or descending into a crisis from which only the developments of the industrial era would engineer a (transient) reprieve.

A difficulty is that no one country is a satisfactory proxy for the whole of Western Europe, not Denmark, not the Netherlands and not England. Most attention will be paid here to England, which solved or bypassed the ecological problems early. Even in England, historical sources and secondary studies of ecological issues are sketchy: there is not much crossover between work by ecologists and work by historians. The alternative to a patchy sketch would be, however, an *a priori* vision of the past that reflects current prejudices. Modern concerns and favourite topics are likely to be jobbed back in an unhistorical way. Such an approach cannot factor in features that have vanished and does not expect to discover fluctuations against the trend. It is unlikely to provide convincing base-lines from which to assess later ecological change. Without attention to documents or the writings of scholars familiar with the ground or close to the sources, ecological history is too easy to present as an inexorable decline in the numbers and abundance of organisms and a deterioration of every type of habitat.

The early modern economy, where we start, was no Garden of Eden for people or wildlife. That too is not to be doubted. Population densities were low compared with what were to come but since technologies were limited the land was used hard.[viii] Resources were prized. During Tudor times, an undertone of ecological competition was present in disputes between landowning families and within communities — disputes that sometimes turned violent. Here are three small examples: a Wiltshire landowner got into a dangerous argument about fishing rights and rabbits in 1545.[ix] Rivalry for power in Bristol culminated in 1579, when one man's retainers set off a battle by breaking into the rabbit warren of another.[x] And a dispute over rights to graze a little island in the river Kennet was taken right up to a central court in London.[xi]

Consider fuel and energy: English woods were not the pristine natural features as sometimes imagined, and most well-wooded parts of the country have an industrial past. The woods were cut on rotation to provide charcoal for iron works. Bark was stripped for tanning on a substantial scale, given that leather making remained the fourth largest industry by value as late as 1851. The woods often formed parts of a wood-pasture regime and were grazed by livestock, which suppressed much plant growth. Since so many commodities were made of wood, trees were reserved for purposes carefully attuned to the properties of different species at different seasons. Few failed to find a use.

The prime use was as fuel. Poor people scraped together whatever they could for firing and broke the hedges when they could not find enough. When nothing else was available, they burned beech leaves and stubble, and the women and children turned out after storms to gather the sticks blown down from the nests in rookeries. In the large areas where woodland was scarce, such as chalk and limestone uplands and the Midland clay plain, the owners of wood sold bundles of twigs and baskets of chips to their unfortunate neighbours, and the commons were grazed so hard that the gorse grew only four inches high. When not cut for bakers' ovens, it was grazed down by ponies. The Weald, which was not an intrinsically barren district, was known as a 'shorn country'.

In places, 'clats' of cow manure were collected, dried and used for fuel, as if England were India. In 1808, the common field regulations for Great Shelford, Cambridgeshire (hardly a wooded area) stipulated that, 'no person shall take off from any common Cow or Horse dung before the same is quite dry and fit for Fuel, to pay two pence for every offence'.[xii] The frequency of regulations of this type is unclear, and the extent of trade to the Midlands in 'clats' gathered and dried elsewhere seems impossible to determine from local sources. Cowpats might be called, rather pretentiously, a dual-purpose resource, and communities had to decide how to balance soil fertility against heat energy.

By early modern times, the bulk of the land, including woodland, had already come into the hands of a small minority, and this makes the sporadic sources on former practices hard to interpret. It is difficult to be sure whether close 'harvesting' of not-very-palatable foodstuffs and searching for minor sources of fuel reflected the institutional factors that rendered so many people landless and poor or came about because resources were stretched in society as a whole. The opportunity cost of the labour of women and children was low at some seasons of the year, that is to say there may have been little they could do towards contributing to family income than gathering and processing tiny quantities of food and fuel. Similarly, the stipulations in farm leases that an out-going tenant was not to plough up and cash in the fertility of permanent pasture, and the requirements on the tenants of some manors to plant every year useful trees like elms or ashes may signify a general need to husband resources or alternatively may indicate a desire not to see assets appropriated to other people's use.[xiii]

Nor was it only in England that ecosystems were hard pressed to provide food, fuel, clothing and shelter when the available resources and methods still comprised traditional materials, hand labour and age-old techniques. We have no reason to suppose that, just because many plants and animals were intensively exploited, the system had no slack and growth within it was impossible. Despite what might be interpreted as signs of strain, the organic economy had more resilience and potential than is usually thought. The picture was far from uniform geographically or chronologically. The variety of ecosystems was great, hampering generalisations, and the economy did not always run smoothly. It had booms and slumps when behaviour would vary, and vary in ways that patchy historical sources do not make clear.

Despite the fact that woodland was valued for so many competing ends, Warde has pointed out that, in southern England, energy was always cheap by continental standards, signifying either that the traditional economy was inherently efficient or that in the centuries before 1800 it actually achieved an energy revolution.[xiv] The latter is a live possibility since, as we shall shortly see, there was a remarkable increase in the productivity of the land before the classic industrial revolution. Paradoxically, then, although early modern England was a place where limitations of technology, lack of man-made materials and the restricted territory in which the average family lived all guaranteed that the environment would be intensively exploited by modern standards, resource productivity was already greatly improving as the industrial era approached — and before it arrived. Mini-crises, small famines, there were, but they can often be attributed to transport and communications too poor to offset every local scarcity, besides the fact that powerful people were able to exclude many at the bottom of the social heap from the resources that did exist. No general crisis is to be found, nor any tendency towards one.

The English farm sector's increasing output of non-foodstuffs confirms how viable agriculture was. An illustration comes from a study by Wrigley that flies in the face of Kjaergaard's depressing picture.[xv] Wrigley wrote to refute another thesis, that of Pomeranz's *The Great Divergence,* which argued that Europe, including England, possessed no developmental advantage over China in 1800, and succeeded in industrializing solely because of its luck in finding coal and seizing the resources of the Americas. Wrigley's counter is the more powerful since he was the scholar above all who emphasized the overwhelming power of the coal economy. He does not recant but does modify his previous thesis. He notes that agricultural historians have concentrated too single-mindedly on the production of food and have grossly underrated farming's role in supplying industrial raw materials and energy for use outside the farm sector. His estimate is that the land's supply of products other than food rose six-fold in England between 1300 and 1800, mostly after 1600. This contribution was twice or three times greater than the growth of population, signalling a mighty increase in farm productivity and in the ability of agriculture to release labour to industry.

It is a reasonable surmise that the energy sector attained similar heights, as Warde suggested.

What happened in agriculture is perhaps not surprising, given the diffusion of new crops and methods. Admittedly, English topography, soils and proximity to markets were so varied that it took a long time for the effects to become general. The innovations may, however, have rescued farming, or the farming of some districts, from declining fertility. Declining or not, low fertility was the bane of European agriculture and a remedy for it was the holy grail endlessly sought. 'As for the Common way and practise in Husbandrie used at this day', Gabriel Plattes had written in 1639, 'all men of good understanding doe know, that it produceth every yeare barrennesse more and more, and in the end, will produce nothing but povertie and beggarie... '[xvi] This touches on the ceaseless debate about agricultural change which finds historians of every period from the Middle Ages claiming that their chosen time saw the decisive improvement in crops and methods. One implication is that neither crisis nor theories of perpetual decline much impress specialists in English or Western European agricultural history: they do not see stasis but instead see continual change, although at an irregular pace.

Agricultural innovation came in rapidly after Plattes' day, or perhaps one should say, far more rapidly than before. All told, there is every sign of poverty before and after his time, but no sign of an ecological crisis in which by 1800 English (and European) resources were stretched to the limit. Further evidence of this, entirely compatible with a growth of poverty, comes from the way in which England, at any rate, could 'afford' to abstract resources from maximum production and devote them to the pleasures of the rich. Over half of the species of tree first planted in Britain between 1670 and 1790 were ornamental rather than utilitarian. Game-bird shooting on landed estates and fox hunting across growing crops interfered with output, even if the organised hunts of Victorian times did eventually offer compensation to farmers. Sport fishing ousted the poor's age-old practice of fishing for the table once the rivers were privatised and devoted to the stylised pastime of fly-fishing for trout. Another sign that old-style resources were ceasing to be vital, at least in the eyes of the well-to-do, was the abandonment of the carp ponds, rabbit warrens and dovecotes that had been part of the manorial economy in the late medieval and early modern periods. During the eighteenth century, fish ponds became ornamental lakes, dovecotes became adornments of gentlemen's residences rather than useful sources of protein in the form of squabs (young pigeons) and when cereal prices were at their height, the rabbit warrens were ploughed up. These trends are no sign that the system as a whole actually lacked resources — quite the opposite.

An Appropriate Baseline

Accounts of past wildlife are rarely written by historians. The accounts are usually quite general and seldom attuned to the shifts in agricultural history from one period to another, or to the diverse geological, climatic and land-holding arrangements that make generalising hazardous. Nevertheless, if there is one lesson that readings of ecology and natural history teach, it is that environmental change had innumerable cross-currents. The responses of animal populations to changes known and unknown are motley and often unexpected. Other than at the most general level, projecting responses from habitat changes is a dangerous game; the effects may seem obvious but species differ too much in genetic adaptability and plasticity to respond automatically in the ways that we may anticipate. If — despite these caveats — we do risk sketching a hypothetical baseline in southern England during the seventeenth and early eighteenth centuries, we shall find that the countryside does not appear to have been enormously rich in wildlife compared with later periods.

Most of our ancestors were materially very poor. Even those who were comfortably placed consumed natural produce to an extent inconceivable now. The raw materials of industry were largely organic and were supplied by the farmer when they were not simply gathered in the fields and woods. The products of the land were used for food, medicine, clothing, transportation, heating, lighting, adornment, entertainment, construction, repair and every household purpose. Modern people are too detached from the land to be able to guess at all the plants and animals that country people harvested until two or three generations ago. This is embarrassingly obvious: natural history is no longer taught in primary schools and I have been in the field with university students who could not tell wheat from barley.

Older books on the countryside make all this plain, they are not mere folklore but repositories of information about the true basis of a hard-scrabble economy. I once asked a Wiltshire farmer how his ancestors had managed on the holding where he lived and on which his family had lived for generations. Speaking of a period as recent as Edwardian times, he answered, 'they lived on what we waste'. Before factory-made goods had begun to spread, the land was used even harder. At a guess, the third quarter of the eighteenth century may have been the worst of times for wildlife. At that time, the population was starting to rise but the changes that were eventually to raise living standards in the countryside were still in their infancy: They include parliamentary enclosures, the ploughing of permanent pasture on the downs and wolds, cheap coal for domestic heating, an exodus to the towns and factories and supplies of industrial goods sent in return.

At that period, and for a long time before and after, the intensity of land use was necessarily high.[xvii] It was the new circumstances of late Victorian times, when exploitation of the countryside began to ease during the depression and the professional middle classes had high incomes and plenty of leisure, which produced the richest accounts of natural history. We tend to think of the past as one vast, lost heritage but our impression of its riches may be an artefact of special conditions that lasted no more than about 70 years, from 1870 to 1940. Two generations of rural decay meant an improving environment for the naturalist! In the third chapter, I shall demonstrate this from what we know of butterfly populations.

Previous conditions had been far less healthy for fauna or flora. During the seventeenth century, huge efforts had been made to drain fenland and convert it to farms. In many Midlands parishes, there were few hedgerows until they were planted to divide the open fields. Studies of the surviving open-field village at Laxton, Nottinghamshire, thus detect low numbers of nesting hedgerow birds. Over much of the Midland Plain, there had been few sizeable woods for almost 1000 years, until game plantations and fox coverts were planted in the nineteenth century.

Elsewhere in England, the typical early modern pattern was wood-pasture, as it had been in medieval times, with livestock grazing amidst a scatter of woods that were intensively exploited for timber, firewood and all the myriad purposes to which wood was then put. Wood-pasture was a poor habitat for butterflies; it was too heavily browsed and grazed, the way woodland is scoured by the excessive deer populations of the early twenty-first century. Wood-pasture and medieval and early modern woodland offered thin pickings for plants and animals associated with old trees, whose only value was for firewood (which meant they were soon cleared away), and for the minor use of tinder when striking lights.[xviii] The downs and wolds were closely cropped by flocks of sheep kept primarily for their wool: Australia had not been discovered. Sheep downs carried an entrancing birdlife but it was a sparse one, restricted to a limited range of species and smaller numbers than after they were put down to cereals. Meanwhile, for almost three centuries, every parish paid a small bounty for the heads of vermin. The birds and animals defined as pests were pursued relentlessly and, although most bounced back when the pressure was lifted, a few were rendered regionally extinct and the populations of others were repressed for a very long time.

Too vivid an impression of wildlife in the past is received from the records the Victorians supplied. Leisured observers became far more numerous and competitive during that reign. The records they have left are heavily weighted towards large and conspicuous species, birds especially, which caught the eye of shooters, or they dwell on butterflies that in a few cases collectors actually drove to extinction. Both birds and butterflies were ardently collected in an age when a specimen was needed to clinch the record: as the saying went, 'it's not a rough-legged buzzard until it's dead'.

How may this be summarised? It amounts to suggesting that early modern England was less hospitable to nature than we might suppose, perhaps overall as well as with respect to species that today's naturalists would most enjoy. Once we refrain from projecting Victorian glories back to preceding times, the early modern period looks to have been a period of such intensive land use that we cannot expect wildlife to have been superabundant. By contrast, the depressions in arable farming between the 1870s and 1940 were in certain respects ecologically blessed. The depressions were interrupted only by the spurt of reclamation during the First World War and might be lumped together as an exceptional era.

There are poignant descriptions of the depression's effects in Eric Ennion's *Adventurer's Fen*, where he laments the loss of wild nature during the Second World War, when the fens were once more drained and brought back under cultivation.[xix] The 'long arable depression' ensured that investment in arable farming was at an abysmally low ebb. Hedges became overgrown, fields were often full of weeds, ditches became clogged and drainage was neglected. Descriptions of the hand labour needed to reverse the neglect during the Second World War make it clear how much had to be done.[xx] Shrinking sheep numbers cut the demand for hurdles within which the flocks might be penned and as a result the practice of coppicing (cutting hazel wands on rotation to make the hurdles) began to waver — spreading the depression from the fields to the woods.

It would seem that bad times for farmers were good ones for nature but the early decline in coppicing reveals how complicated the matter is: species associated with the high woods and closed canopies may have gained but those that prefer sunny glades, most conspicuously certain butterflies, suffered when the coppice openings were no longer cleared every 7 or 14 years. Subjective judgements have to be made as to which version of 'nature' we prefer and are willing to pay for, however much talk there may be of a scientific — meaning merely natural science — basis for conservation.

Looking more closely than is usual in the natural history books, we can see that wildlife is systematically affected by the cycles of the rural economy. It is contingent: ecology is substantially a function of economic life. When should the baseline be set against which to measure the status of the flora and fauna? The norm in early modern England was no crisis but nevertheless was a leaner time for wildlife than the regime that came into being between 1870 and 1940. During and after the Second World War, agriculture was revived. Wartime food shortages, and the post-war subsidies that governments found too hard to remove in the teeth of incessant politicking by the farm lobby, brought losses of ecologically important sites and declines in many species. SSSIs (Sites of Special Scientific Interest) had been designated after the war but were never given sufficient legal protection. Farmers ploughed up numbers without penalty. From the naturalist's point of view, this was a scandal but it refers to the circumstances of a given epoch. There is no warrant for thinking of the whole past in terms of irreversible decline. Whatever is happening now and whatever may happen in the future, it is clarifying to see from its history how far the environment has been and must remain an epiphenomenon of the economy.

ENDNOTES

[i] T. Kjaergaard, 'Denmark's Ecological Crisis in the Eighteenth-century,' in *Economia e Energia Secc. XIII-XVIII* (Florence: Le Monnier, 2003), pp. 905–915.

[ii] V. Barbour, *Capitalism in Amsterdam in the Seventeenth Century* (Ann Arbor, MI, USA: University of Michigan Press, 1963), pp. 46–47.

[iii] T. C. Smout, *Nature Contested: Environmental History in Scotland and Northern England since 1600* (Edinburgh: Edinburgh University Press, 2000), p. 65.

[iv] *Ibid.*, p. 68.

[v] T. Kjaergaard, 'A Plant that Changed the World: The Rise and Fall of Clover 1000–2000,' *Landscape Research*, **28** (1) (2003), pp. 41–49.

[vi] A. Geim, *Financial Times*, 3 July 2012.

[vii] V. Smil, 'Population Growth and Nitrogen: An Exploration of a Critical Existential Link,' *Population and Development Review*, **17** (4) (1991), pp. 569–601.

[viii] E. L. Jones, *Locating the Industrial Revolution* (Singapore: World Scientific, 2010), Chapter 3.

[ix] F. E. Warneford (ed.), *Star Chamber Suits of John and Thomas Warneford* (Trowbridge: Wiltshire Record Society, XLVIII, 1992). One of those engaged in this dispute, Rauf [Ralph] Cawle, was perhaps a distant ancestor of mine.

[x] Joseph Bettey, 'Feuding Gentry and an Affray on College Green, Bristol,' *Transactions of the Bristol and Gloucestershire Archaeological Society*, **122** (2004), p. 154.

[xi] Hungerford Local History Group, *Elizabethan Hungerford* (Hungerford: privately printed, 1995), p. 11.

[xii] See Sites.google.com/.../greatshelfordhistory.

[xiii] For example, *Victoria County History, Gloucestershire*, VII (Oxford: Oxford University Press, 1981), p. 23.

[xiv] P. Warde, 'Woodland fuel, demand and supply', in J. Langton and G. Jones (eds.), *Forests and Chases of England & Wales, c.1500–c.1850* (Oxford: Oxbow Books), p. 82.

[xv] E. A. Wrigley, 'The Transition to an Advanced Organic Economy: Half a Millennium of English Agriculture,' *Economic History Review*, **LIX** (3) (2006), pp. 435–480.

[xvi] G. Plattes, *A Discovery of Infinite Treasure Hidden Since the World's Beginning* (Norwood, NJ, USA: Walter J. Johnson, 1974 edn. [first, London, 1639]), p. 17.

[xvii] E. L. Jones, *Locating the Industrial Revolution* (Singapore: World Scientific, 2010), Chapter 3.

[xviii] P. S. Savill *et al.*, *Wytham Woods: Oxford's Ecological Laboratory* (Oxford: Oxford University Press, 2010), p. 33.

[xix] E. Ennion, *Adventurer's Fen* (London: Methuen, 1942).

[xx] E. Blishen, *A Cack-handed War* (London: Thames & Hudson, 1972).

2. DECLINE

> We might have worse evidence of the effect of agricultural improvement than the fact that in two years (1710–12) which Professor Ashton puts 'among the worst of the (eighteenth) century' no one wanted to hang the graziers or have the enclosers in the ditch.
>
> Maurice Beresford, *Habitation versus Improvement* (1961)

History was long thought to be directed, maybe determined, by forces external to humanity. The Ancient Greeks attributed the waywardness of life to the maliciousness of the gods. Christians hold public prayers asking God to intercede during floods or drought. They have done so recently, even in Western countries, accompanied by opportunistic pronouncements that some disaster or other is heaven's retribution for human sin. When they are not blaming homosexuality or women bishops, priests and laymen alike are inclined to claim that the sin lies in practising an economic system of which they disapprove. This is invariably a (chronically underspecified) capitalist market system of which the critics are themselves the beneficiaries; the appalling environmental record of command systems, so much harder to correct, rarely comes in for criticism. Alternatively, human activity as a whole gets the blame for supposed environmental degeneracy.

One term used is 'landscape decay', and when the subject is the old-settled area around the Mediterranean, the finger is pointed straight at human misuse.[i] The misanthropy which finds humanity a blot on the southern European landscape turns into 'ruined landscape theory'. This proposes that exploitation has turned the Mediterranean lands into a one-way street of damage, depopulated because the soils and vegetation have been over-used. In their work on *Mediterranean Europe*, Groves and Rackham reject this jaundiced interpretation.[ii] They show that the empty landscapes around the Mediterranean were abandoned by choice. People were not driven out by environmental degradation but moved away voluntarily in search of better jobs. There was no monotonic decline. Groves and Rackham bring out the historical contingency of landscape types well and dispose of other common failings in works on ecology. The failings include neglect of the social gain, which is as though humanity is the only species deserving no consideration whatsoever.

Texts about the environment, some written by professional ecologists, are inclined to give the impression that the Earth was once in a pure, pristine condition — a Garden of Eden before the Fall — from which changes wrought by humans have always been for the worse. This 'declinist' view is sometimes conveyed by winks and nods rather than direct statements, but the meaning is usually plain. The most influential formulation is by the Dutch ecologist, Frans Vera, who borrowed from marine ecology the concept of a shifting baseline.[iii] His thesis is that every cohort of conservationists starts at a baseline that is already a step down from previous times; the stairs continue downwards and no-one remarks how much higher they were just before their own day. His alternative metaphor is of a carpet being cut into pieces, smaller and smaller and ever more adulterated by human exploitation and invasive species. Each fragment contains fewer and fewer of the previous populations of plants and animals (notice the implication that newer entrants are of lesser value). By any earlier standard, the new baseline is always more impoverished, as if humanity had invented the most dismal of perpetual motion machines.

Vera's inspiration has led to experiments with a depopulated grassland and wetland at Oostvaardersplassen in the Netherlands. This former polder abuts the Ijsselmeer just east of Amsterdam and teems with birds and animals, leading a visitor to wonder how many more could possibly have been accommodated in the past. Large herds of Red Deer, Konik horses and quasi-Aurochs cattle have been introduced in order to track the effects of long-term grazing.

Notwithstanding the enviable abundance of wildlife on this polder, continual historical decline is not wholly implausible. It chimes with common understanding of ecological change through time, although we should be wary: the retrospective accounts of nature that form our implicit baseline always tend to sound lyrical. Leaving to one side the question of what previous generations may realistically be expected to have done when faced with both their own poverty and the fragmentation of habitats, the problems of conservation lie less with the Oostvaardersplassen experiment than with matters that it does not take into account. By assuming that ultimate value resides in the original ecosystem, the approach is loaded from the start. Gaps may therefore be thought to exist in the argument underlying the experiment. Let us consider them in turn, bearing in mind that I am perhaps unfairly using the following few pages to criticise approaches to ecology broader than just that at Oostvaardersplassen.

First, the implicit goal seems to be the 'original' natural ecosystem, with no cherishing of subsequent stages of development or recognition that humanised landscapes may be as attractive. Nature frames the whole experiment, and consideration is not given to the values of human society. Yet do segments of 'pure' habitat like those meant to be recreated at Oostvaardersplassen truly maximise human preferences? It is not a pure ecosystem because there is intervention on animal welfare grounds. Large animals dying in full view are deemed unacceptable to the general public. A degree of culling is carried out, which disturbs the pristine ecological intention that was first in mind. Curiously, I am told that while people react badly to the sight of beasts starving or dying in the winter when this is posted on YouTube, they do not seem to notice carcases left to moulder on site — as they were on one April day when I visited.

Since taxpayers are funding the experiment, and other uses of the land have to be forsworn for it to happen, it is fair to ask whether a majority of people might not prefer the restoration of something resembling the landscape of Holland at another period, say one depicted by the Dutch Masters. Alternative historical and ecological recreations would be possible and a different type of experiment could take place. Recreating humanised landscapes, complete with the 'living entourages' of their day, is surely as likely to satisfy the curiosity of the public as an ecosystem whose purity is already sullied by politically correct objections to leaving large species to die, so to speak, on camera.[iv] Admittedly the problem would arise of which periods to select. Rembrandt's Holland? Yet maximum biodiversity is a curious goal, with so many cross-cutting options and so much competition among groups of species that it is hard to define. Why should maximising be the goal? It is as if landscapes that incorporate historically known types of plant and animal communities are, in some obscure sense, less desirable than an ill-defined original maximum.

A second feature of many proposals for conservation is that private property rights are ignored, i.e., the legal right of individuals to use their land as intensively as the market indicates is overruled. Compulsory purchase and command-and-control solutions are preferred. On the contrary, some may think that producing food in a world where a billion people go to bed hungry is *ipso facto* desirable; others may think the rule of law is intrinsically important. It is not necessary to adopt either position to notice that the issues are not discussed.

Thirdly, just as historical endpoints are slighted, so are historical processes. Maybe the overall trend really has been for biodiversity to decline from one equilibrium to another, but no allowance is made for counter trends — periods of upward fluctuation in species variety and density. Nor is the rationale of the trade-off between biomass and diversity always clear. Fluctuating intensities of resource use in recent centuries mean that care needs to be taken over the baselines selected, an issue which this book will mention repeatedly.

Biomass and biodiversity have altered under different forms of human occupation, not always in parallel and not always, as it were, trending downhill. A different example from that of the Oostvaardersplassen will serve to expose the value judgements that are unavoidable when discussing historical change. Let us therefore make a long excursion to the chalk downs of southern England. It is commonplace for naturalists to wax nostalgic about the centuries when these downs were covered with velvety turf, close-cropped by sheep and roamed by the great bustard and the thin numbers of other evocative species recorded in the seventeenth century by the Wiltshire antiquary, John Aubrey. The arable depressions of the late nineteenth century and between the two world wars (entirely inadvertently) began to reproduce Aubrey's seventeenth-century downland. Cereal growing went on the retreat. Some extensive grass-fed sheep enterprises re-emerged. The hamlet of Snap in Wiltshire was depopulated in favour of sheep grazing, as if by some overbearing Tudor encloser. Fields of grain were replaced by tumbledown grass, sheets of St John's Wort, thorn savanna full of nesting Red-backed Shrikes, and flocks of Stone Curlews in the autumn. The value of pheasant and partridge shooting was greater than the farm rents, if farmers could pay rent at all. Compared with the dull plough land and cereal monoculture of the post-war years all this seems romantic, and as a child of the post-war period, I was by no means immune from day-dreaming about it.

Yet the inventory of nature can be only part of the story. The ruinous agriculture between the wars levied a serious human cost. As my father told me (we lived on the Hampshire chalk), farmers came into town with their cars tied up with binder twine, farm buildings were crumbling, cottages were neglected and sometimes deserted and labourers were on the dole. A preference for the natural history associated with this regime is a matter of taste, but maybe a narrow, even a selfish, taste.

The weedy, tumbledown fields of the interwar years contained relatively few bird species in rather low numbers, just as the grassy downs did in the seventeenth century. Who is to decide they were preferable to the greater biomass of the post-war arable fields, just because downland grass became scarce as a result of the wartime and post-war ploughing campaigns? Who is to decide that birds of the downland are more to be admired or of more interest than the flocks of seed-eaters that replaced them when the chalklands became arable? Counting small birds on winter ploughland is a real challenge; they are hard to find among the furrows. The task is not obviously easier than seeking grassland birds, and although 'twitchers' seldom check out the flocks, doing so is not obviously less fun than their obsessive quest for rarities.

Under any given farming regime, so many other organisms are present, often small and unseen, that biological comparisons between periods are hard to make. Naturalists may yearn for the past but most of them seem to ignore the context and place the Golden Age either when they were young or just before that glorious dawn. They over-represent in their minds' eyes the more conspicuous groups of organisms like birds (*mea culpa*). Anything that has vanished from the mix is considered cause for lament, whereas introduced or invasive species (think Collared Dove) are not much prized. The losses may remind us of Vera's descending escalator but this misses the point. Ecosystems accumulate gains as well as suffer losses, striking a balance between them is partly subjective, and deciding on a baseline is as difficult as it is necessary.

VERY LONG-TERM HISTORY

Does the history of the human use of the earth bear out the declinism thesis? In a very general sense, the answer is yes, although we should take care to avoid the automatic 'man is evil' presupposition and bear in mind just how limited the data are. The breadth of the estimates of land use reinforces the latter point; Goldenwijk, for instance, asserts that from 15% to 25% of forested land worldwide, and an even larger area of natural grassland have been cleared since A.D. 1700.[v] One-third of the global land surface is now farmed. The fragility of the estimates is obvious and in the absence of documentation they can never be precise. Presumably reaching this point involved massive losses across the plant and animal spectra. The Nile valley, to choose a key location, was transformed in ancient times by imported plants, and the land was irrigated to grow more crops.[vi] Desertification in the vicinity is said to have been driving people into the valley; the possibility that they were actually attracted by better opportunities scarcely gets a mention. Elsewhere, the world's arable land remained for millennia a series of islands in a vast expanse of forest and grassy plains.

The long, slow evolution of farming systems escapes most types of record, though occasionally a relatively rapid supply surge can be detected.[vii] One of these episodes was the introduction of Asian crops to sub-Saharan Africa, where they spread very gradually. The 'Arab Agricultural Revolution' brought a number of food crops west from India into the Middle East, North Africa and Southern Europe after the eighth century A.D. Another early transfer was of early-ripening Champa rice from Vietnam into China. Accounts of these shifts have however become less dramatic in the light of later research. Less controversial, and on a much grander scale, was the 'Columbian Exchange', which spread crops from the Americas around the world. Whereas it would be possible to argue that the previous changes had largely replaced existing crops with higher-yielding varieties, producing no great net effect on the landscape or native animal and plant species, it is known that the dry-land crops which were part of the Columbian Exchange contributed to a great extension of cultivation in China. The Han Chinese pushed southwards over the centuries — one of many folk migrations in history — and cleared some 670 million acres of forest. There are estimates of sizeable increases in the arable areas of Japan and India too, and it is impossible that there were not vast consequences for flora and fauna.

Even so, we should still resist a unilaterally declinist position. Laurence points out how hard tropical biodiversity has been hit by the combined effects of selective logging, fires, habitat fragmentation and the hunting that has been facilitated by logging roads.[viii] Quite so, but he also notes that, although several species of birds and primates have decreased, lemurs and many butterflies have increased, thanks to more ground cover, foliage and fruit. The mix of tree species alters according to their relative light tolerance and their overall variety has not tended to decrease. Populations of light tolerant insects have expanded. Laurence warns that the eventual effects of repeated cycles of logging are unknown, yet this would not justify invoking the 'living dead' argument, that is to say, implying that particular ecosystems are inherently doomed despite the lack of current evidence. Why should the effects of repeated logging differ in principle from those of the shifting cultivation characteristic of tropical history?

Laurence is concerned with the recent past. Returning to the longer term, we should note, too, the limitations of Goldenwijk's bold attempt at sketching global land-use history. He uses historical population density as a proxy for former agricultural activity and introduces many estimates of land-use change which he terms 'data sets'. They are not data; they are conjectures. He does, however, usefully challenge the assumption that cropland has always been located where it is now found. The total area has risen but this may be misleading, given the 'forest transition' which has reversed the retreat of woodland in many parts of the world. Farming is concentrated today on the better soils, and rural people are shifting en masse to cities, quitting their fields in the way that land around the Mediterranean was abandoned. Absolute declines in wildlife cannot always be assumed when reversals or partial reversals are constantly taking place.

In sum, we need tests of the proposition that nature today is nothing more than an impoverished step low on an ever-descending staircase of environmental history, each age seeing smaller, more fragmented habitats and less abundant wildlife than its predecessor. The thesis should apply to any group of organisms in any countryside, yet cases well documented over any length of time are not easy to find. Environmental books at large, books on environmental economics in particular and even some histories of the environment, often show little familiarity with natural history or the species whose conservation is devoutly desired. Their scope is general, their particularity limited and their attention to documented trends is notable for its rarity. In the next chapter, we shall test the thesis of relentless decline by looking at episodes in the experience of the butterfly fauna in England. Prepare, therefore, for an abrupt shift from grand schema to close-up view.

ENDNOTES

[i] B. de Vries and J. Goudsblom (eds.), *Mappae Mundi: Humans and their Habitats in a Long-Term Socio-Ecological Perspective* (Amsterdam: Amsterdam University Press, 2002), p. 48.

[ii] A. T. Grove and O. Rackham, *The Nature of Mediterranean Europe: An Ecological History* (New Haven: Yale University Press, 2001), pp. 10–17.

[iii] For example, F. Vera, 'Re-Wilding — The Dutch. Experience,' Report, British Ecological Society Meeting, 12 and 13 July 2007. Available at: www.wildland-network.org.uk.

[iv] The issue of cruelty is grappled with in the second (online) report of the International Commission on Management of the Oostvaardersplassen, November 2010.

[v] K. K. Goldenwijk, 'Footprints from the Past: Blueprint for the Future,' in R. DeFries *et al.* (eds.), *Ecosystems and Land Use Change* (Washington D.C.: American Geophysical Union, 2004), pp. 203–215.

[vi] 'Tuinen van de faraos' (Gardens of the Pharaohs). Exhibition in the Rijksmuseum, Leiden, April 2012.

[vii] E. Jones, *The Record of Global Economic Development* (Cheltenham: Edward Elgar, 2002), pp. 50–59.

[viii] W. F. Laurence, 'Rapid Land-Use Change and its Impacts on Tropical Biodiversity,' in R. DeFries *et al.* (eds.), *Ecosystems and Land Use Change* (Washington D.C.: American Geophysical Union, 2004), pp. 189–200.

3. PROPER BASELINES: THE EXAMPLE OF ENGLISH BUTTERFLIES

Even the descriptions of butterfly numbers during poor summers force our occasional years of relative plenty to pale into insignificance.

Matthew Oates (1996) on the New Forest in Victorian times

A study of English butterflies will enable us to dig into historical fluctuations in the natural world and test the thesis of sequential decline. It will also permit us to take a critical look at the way recorded history is treated in the biological literature. Butterflies have long been an object of interest to enthusiastic amateurs, so there is a large literature based on extensive field observations to examine. As a former Director of the Edward Grey Institute of Field Ornithology said with respect to bird-watchers, amateurs may be amateur but they do gather information beyond the resources of professionals to collect. We can plunge into the specialised reaches of butterfly collecting and regional natural history to see if every period really was entomologically poorer than the one before.

The question is as follows: is the present distribution of English butterflies only the threatened remnant of abundance present in former times — although on a successively diminishing scale — until our own degenerate age? This is the sense conveyed by the writings of lepidopterists. During the nineteenth century, Shannon tells us in *The Aurelian Legacy*, butterflies were extraordinarily profuse. He rests this conclusion, not unreasonably, on descriptions of the more varied rural land use combined with ecstatic reports by Victorian butterfly collectors. Miles of close-cropped sheep walks were then alive with adonis and chalkhill Blues, while 'every' parish contained ancient woods, managed as coppice, containing fritillaries, white admirals and hairstreaks.[i] Coppicing meant that, every 7 or 14 years, hazel woods were reduced to the basic 'stools' or stumps, surrounded only by a low ground flora. Glades were opened and the light let in. The decline of light-tolerant butterflies is commonly attributed to the reduction of coppice, of which England contained 134,000 hectares as late as 1947 but only 22,000 hectares by 1998.[ii] The decrease of active coppicing is obvious on the ground, where old, untended stools are frequent.

To Shannon, the nineteenth-century countryside had been a patchwork of attractive land uses and, as late as 1900, there were plenty of permanent pastures bordered by thick hedges and harbouring ponds in their corners. He quotes J. B. Stephens's description of collecting in Surrey in the 1820s, when along miles of hedgerow myriad white-letter hairstreaks hovered over every flower and bramble blossom. 'Some notion of their numbers may be formed', Stephens wrote, 'when I mention that I captured without moving from the spot, nearly 200 specimens in less than half an hour.' Stephens did have the grace to admit that this 'exceeded anything of the kind that I have ever witnessed'.

There was many a rapturous account in Victorian times. Shannon quotes F. W. Frohawk as remembering the silver-washed fritillaries of the New Forest, Hampshire. In the latter years of the nineteenth century, according to Frohawk, 'it was common to see forty or more assembled on the blossoms of a large bramble bush, in company with many white admirals, meadow browns, ringlets... '[iii] And Oates too quotes Frohawk as describing a day in the New Forest in July 1888, when, 'insects of various kinds literally swarmed. Butterflies were in profusion... '[iv] Oates adds a remark by another collector about a visit in 1892 when butterflies of several species, 'were so thick that I could hardly see ahead and indeed resembled a fall of brown leaves', while the next year, 1893, a different collector found the bed of a stream crowded with butterflies for over a mile. Abundance like this no longer characterises the poor, battered New Forest.

Observations of this type need not be doubted. As far as collectors were concerned, the countryside of Victorian England — made accessible by the steam train — was a Paradise. What may be pondered is whether their accounts strike a fair average of the deep past. What may the situation have been like before their day? Many authorities convey the impression that, before the decay of modern times, England was always a Garden of Eden. But was this true?

Shannon's argument is that, in the nineteenth century, there were no insecticides or artificial fertilisers (which is not quite correct), or any motorways or other signs of the development current at the turn of the twenty-first century. Every change since the nineteenth century is made to seem gloom and doom. His view is shared by other lepidopterists. 'Even our common butterflies have become severely reduced in number', declares Tomlinson, 'and, for many people, butterflies are no longer an intrinsic part of summer days'.[v] Given the colourful and widely admired red admirals, peacocks, commas and so forth present on buddleia, a frequent bush in England's ubiquitous gardens, this negativism seems a little overwrought. Tomlinson states that the loss of butterflies 'is symptomatic of the exploitation of the environment that has occurred and is continuing.' If butterflies are all one is interested in, deprecating human activity may be understandable, but that is scarcely the attitude of a responsible citizen.

As far as Tomlinson's version of the losses goes, it may be so, although his species accounts add up to 17 increases and 17 declines, which is scarcely conclusive. Detailed survey data exist only for the very recent period, 1976 to 2011, but their interpretation varies according to which species we consider. If we take species for which the trend is statistically significant and the sites surveyed for each are numerous (over 50), those showing increases are marginally commoner than those showing declines (seven against six)![vi] Losses are most heavily concentrated among so-called specialist species, i.e., woodland ones, where the causes are likely to be habitat changes resulting from reduced coppicing and heavy browsing by deer. Generalist species apparently show no particular trend.

Whereas some species decline and their ranges retreat, others increase and expand, for example, the large and small whites characteristic of cultivated land. The typical amateur lepidopterist is unwilling to spend his or her time studying these equivalents of the weeds of cultivation. Scarce species are those that excite the observer; and they are the ones most likely to be at risk from changes in farming, forestry or commerce. But it seems unscientific to refer to the costs of exploiting the land and not the gains. As I shall repeatedly observe, the economic growth blamed for losses is precisely what enables more people than ever to indulge a taste for the natural world and provides the funds for the conservation movement. The point is surely understood by most people but an exclusive concern for other species obscures our own interests, which are placed outside nature.

Some entomologists put themselves in the camp of the true 'people-haters' by abhorring modern development. Take E. B. Ford's classic *Butterflies* (1945), in which he states that since the start of the nineteenth century, 'urbanisation and industrialism have made vast and hideous advances'.[vii] He adds the politically contentious remark that the great estates, 'so important to rural communities and to the preservation of wild life, are collapsing under insupportable taxation, and the country will ultimately be infinitely the poorer for their loss.' Still more contentiously he refers to the horrors of urbanisation and industrialism in Oxford (in 1943!), by which I assume he means the industrial suburb of Cowley. At the risk of an *ad hominem* argument, it was easy for someone with a privileged life like Ford's — he died in his bed at All Souls at 86 — to show no sympathy for people who lived in terrace houses and made their living in factories. As the entry for him in the *Dictionary of National Biography* pointedly remarks, he knew nothing about the political world and cared less.

Ford may have been extreme in expressing his views but he was, and is, far from alone in his attitudes: the endemic NIMBY resistance to building development shows that.[viii] Yet why should we lament what happens to certain species when biologically speaking others are equally interesting — for instance, 'pests' are interesting as well as important. As Jeremy Thomas says of butterflies, the large white — the cabbage white — is one of the most interesting butterflies in Europe because of its parasites, the use it makes of mustard oils and its remarkable migratory flights.[ix] The botanist, Edgar Anderson, long ago noted the prevalence among naturalists of the 'cloud forest syndrome', where attention is concentrated on rarities to the exclusion of the weeds — and we may add cabbage whites — that are common around settlements.[x] Anderson's point deserves frequent repetition. It is scarcely an efficient use of research funds to rush off to mountain tops when so much may be learned more cheaply on the doorstep.

Why, too, do writings on ecology treat our own species as noxious or irrelevant? Admittedly, the damage 'we' did to the English environment in the twentieth century by intensive farming and its creation of monotonous habitats was needless; the country could have imported food at a lower resource cost than growing it at home. We could have had our cake and eaten it too. Yet when deprecating trends like the reduction of habitats for butterflies, surprisingly few pertinent sources — the Millennium *Atlas of Butterflies* is an honourable exception — mention the transfer payments from taxpayers and consumers to the unquenchably demanding farm and landowning lobbies.[xi] The emphasis in most butterfly books is on environmental consequences rather than causes. The switch that has lately been made to a single farm payment whether food is grown or not scarcely seems to have produced environmental gain, except perhaps where field margins are left rough (which is itself a subsidy to the rearers of game birds). Ford, who thought pheasants were the most deadly scourge of woodland butterflies, would not have approved, although he does not say whether the effect he deprecates came from the birds eating butterfly larvae or because game preservers cleared away brambles and undergrowth.[xii]

The entomological literature is full of statements along the lines that butterflies constitute the longest running indicator of the environmental condition and are more representative than birds. The arguments are that butterflies occur in all the main terrestrial habitats in the United Kingdom, and because insects constitute a large percentage of terrestrial wildlife, their trends must mirror those of the whole. This is contrary to the common assumption that problems will reveal themselves first at the top of the food chain, as the negative effects of DDT did when peregrine numbers crashed — a decrease reversed by bans on the indiscriminate use of DDT, so that peregrines now raise their young on church towers in many an English town. Moreover, the concept of 'environmental condition' is ambiguous because the same habitats do not suit all species equally — another point that deserves to be underlined. Survey lists seem to consist of counts of butterflies set against casual descriptions of habitat. Descriptions of the management status of Forestry Commission woods are surprisingly general and are not scored in a consistent numerical fashion.

Nor are long series of observations brought into play. The trends analysed refer to the brief spell since the 1970s, very occasionally since the 1960s, rarely from as far back as the 1940s. These short spans simply do not incorporate the series from the early eighteenth or late seventeenth centuries to which allusion is continually made. For instance, Pollard and Moss title a paper, 'Historical records of the occurrence of butterflies in Britain (etc.)', but analyse the records of just three species and only from 1900 to 1966.[xiii] In practice, they offer correlations for only two species and then merely for the years from 1976 to 1991.

More satisfactory historical papers may exist but the above reflects the tendency. Amateur entomologists are inclined to view change against two implicit baselines, first a Garden of Eden supposedly characteristic of past time, when butterflies are presumed to have been always more abundant and widely distributed than today, and secondly, conditions remembered as superior either from their childhood or from when they first took up the hobby. Memories tend to be nostalgic, recalling 'good' excursions and overlooking unsuccessful ones, dwelling on particular times of plenty, as well as sorrowing for favourite sites that have been cut down, ploughed up or built over. Backward-looking books on the countryside, a literature in which England excels, reinforce the impression that times gone by were rosier than now; they should be read critically and with attention to the precise period they describe. They seldom point out the advantages in equipment, information and travel that characterise the present day. None of this denies that there have been unfavourable changes but the questions are 'unfavourable for which species' and how representative are the reports and memories.

The most promising historical article is one by Downes on the speckled wood.[xiv] Almost alone among those I have read, it does examine primary historical data, not merely on the species of its title but on seven other British species, comprising 13% of the resident fauna. All eight butterflies underwent considerable expansions or contractions of range at some time during the nineteenth and first half of the twentieth centuries, sometimes both trends at different times. Many alterations were so rapid, Downes notes, that interpreting the distributions to be found at the end of the 1940s by the methods of classical zoogeography would be 'very difficult'. He adds that when change came so fast any suggestion of geological or glaciological causes would be pointless.

The scope of Downes's research justifies his sceptical conclusion. He very properly examined original records, these being the observations of selected species printed in the runs of three leading entomological journals. He studied the files of *The Entomologist, Entomologist's Monthly Magazine* and *Entomological Record* from their inception. Unfortunately the first volumes of these journals date only from 1840, 1864 and 1890, respectively. What they pick up, therefore, are distributions reported by entomologists in the Victorian hey-day of collecting. Research into earlier records could surely double the length of time that might be reviewed.

A marked feature of entomological work in very recent years is a determination to demonstrate the influence of climate change. There is a bias now towards attributing changes in the status of butterflies to alterations in the climate or even in the weather of a few seasons. Habitat changes are less easy to code and by contrast are played down. One element in this *a priori* choice is almost certainly opportunistic: the availability of grants to fund the work. Yet, while the weather may influence the abundance of insects at given sites, if there is no site with suitable food plants, no alteration in the weather is going to preserve it or bring about large extensions of range. Spillovers to colonise marginal areas or food sources may take place if numbers become insupportable at a given site but it is difficult to grasp how distribution may be much affected in the absence of suitable territory. Speculation about the rise of differently adapted races that have managed to cope successfully seems to be just that, speculation, though perhaps studies of DNA may alter this.

A tendency to elevate the importance of climate change has not been deflected by the conclusion of the Millennium *Atlas* that the main cause of butterfly decline has been, 'the unprecedented rate of loss of semi-natural habitats'.[xv] The *Atlas* is undoubtedly correct in blaming this on increased economic activity in the countryside, largely through policies directed at agricultural intensification. The bias of *Atlas* opinion is clear. It refers repeatedly to agricultural 'improvement' in scare quotes, as if this is necessarily illegitimate. In terms of the law of the land and the technical efficiency of farmers, improvement was to be expected, whatever its side-effects. The environmental effects have been deleterious, but such is the democratic choice.

The *Atlas* is written from a conservation standpoint and deprecates any development that tends to reduce butterfly ranges or abundance, to which end it refers to vulnerabilities even when no threat is manifest. This nervousness is found elsewhere: for example, Tomlinson describes species as 'vulnerable' to environmental change when they are currently doing well. This is the 'living dead' gambit: just you wait, it implies, things will get worse. At one point Tomlinson refers to a potential threat to the white admiral from the browsing of honeysuckle by muntjac deer, which he describes as an 'alien species'.[xvi] This betrays a static view of the environment, a dislike of change that surfaces whenever introductions are discussed. A more scientific opinion was expressed by L. Hugh Newman, who urged that fauna and flora always represent cross sections of the processes of change and are proper subjects of study in their own right.[xvii] Why, Newman asked, should all causes be regarded as natural except the actions of humanity? His more genuinely scientific approach is one I shall mention again in later chapters.

A prime difficulty facing butterfly conservation is that the habitat requirements of butterflies often appear at variance with one another. The shaded woods that grow up when coppicing ceases may be satisfactory for white admirals and commas, but not for species that frequented the sunny glades where coppice had been cut. Pearl-bordered fritillaries were once common but now that coppicing is rare, they have become rare as well; to find them in ones and twos, flitting low in their bright orange, it is necessary to tramp about amidst encroaching bracken and silver birch.

Reduced grazing pressure, after the farmers had spread myxymatosis to blind and kill off the rabbits, and after farming had been withdrawn from marginal pastures, favoured species like the duke of Burgundy fritillary, marbled white and Lulworth skipper, all of which prefer long grass. The marsh fritillary is thought to have spread onto chalk down-land during the early twentieth century, especially on Salisbury Plain, when this was requisitioned by the military and arable farming retreated. On the other hand, some blue butterflies were disadvantaged by the coarsening of the turf that sheep had once kept closely cropped.

Two aspects of the variability of response may be emphasized. Firstly, there is a measure of surmise in most accounts and explanations of change in the ranges of butterflies; secondly, we should notice the contingent nature of distributions given (for instance) that natural succession eventually leads ungrazed grassland to return to scrub. There is no unique fixed point; interventions by conservationists must favour the interests of some species rather than others. The issues that come to the fore therefore relate to which species we should prefer and which land uses are to be selected in order to maximise the presence of one or the other. Overarching policies affecting land use and land management cannot satisfy every wish. The intricacy of habitat requirements calls for a case-by-case approach. We are told that our objectives must be clear, 'before conservation can be put on a firm scientific basis'.[xviii] But these are not issues where natural science is competent; they are matters of social or political choice.

The *Atlas* regards modern human activities as necessarily detrimental to the landscape, habitats and species. Ford's classic work is quoted on page 336 as declaring that, 'since the beginning of the nineteenth century the English countryside has become steadily less favourable for butterflies.' This was belied by the evidence even when he wrote in 1945 and is at the very least overgeneralised — compare the note in the *Atlas* (p. 332) that late twentieth-century recolonisations by orange tip, peacock, comma, speckled wood and marbled white demonstrate 'the rapid and dramatic nature of fluctuations in the distributions.' Fifteen species at least are reported to have increased or expanded in recent years and another five species to have shown small recoveries. Cycles of abundance in some, for example the holly blue, take several years to become plain.

A flaw was Ford's use of 'steadily', which carries with it the suggestion, not merely that once upon a time there were simply more butterflies in more places, but that there has been uncompensated deterioration. As a gross generalisation, this is perhaps forgivable — we would all like to see more butterfly diversity — but it obscures too much history. The true extent of fluctuations in density and distribution is disguised and all the influences on the natural world may not be recognised. It also means that the claim, beloved of conservationists and repeated in the *Atlas*, that butterflies are indicators of the (presumably overall) quality of habitats is too broad to be of practical use. The weak correlations between changes in distribution and the explanations offered throw doubt on 'indicator theory'. How can butterflies be universal indicators of the condition of the environment when the influences on the species exhibiting the most dramatic fluctuations of all, the comma, consist of nothing more than inconclusive allusions to the weather and speculations that better adapted races may mysteriously have appeared in the nick of time?[xix]

Is the record as negative as implied? 'With butterflies becoming scarcer every year', as the Collins *Field Guide to the Butterflies of Britain and Europe* stated at the start of the 1970s.[xx] Have there been no changes in a positive direction? Of course there have. An approximate periodization which throws light on this may be compiled from the national overviews by South (1906), Ford (1945) and for woodland species Ford (1948).[xxi] These texts have the merit of having been written in the Victorian era or cite writings from the nineteenth as well as the twentieth century. Although they may be supplemented by such works as the Millennium *Atlas* (2001), Tomlinson (2002) and Thomas and Lewington (2010), tabulation is difficult because the observations are seldom standardised.

The definitive work remains the Millennium *Atlas*. This claims specifically to incorporate material from 1800 and before in its data-set, and shows the results on a distribution map for each species. Unfortunately, as is the case in almost all sources, the promise exceeds the performance: the *Atlas* maps mark early records only as 'pre-1970'! Earlier fluctuations in numbers or range can seldom be picked up, when they can be picked up at all. The detailed surveys made during the late twentieth century distract attention from earlier records which cannot compete in terms of coverage. The prospect of locating historical trends is defeated because early information does not measure up against modern statistics. Attracted nowadays to analysing great bodies of data, biologists seldom bother to trawl through the historical records that are available. They act as if these lesser sources can tell us nothing: the best drives out the good.

Some local effects do not generalise enormously well, notably in the case of the New Forest. In an entrancing paper, Oates deals extensively with nineteenth-century experiences of butterfly collecting and woodland management in the Forest.[xxii] He tabulates observations for sub-periods and displays a picture of ebbs and flows in abundance and distribution, although — conforming to the fashion for telescoping past time — his first column is blandly labelled 'pre-1900'.

Forestry Commission policies seem largely responsible for transforming the once-rich habitats of the New Forest compartments called Inclosures, and the broad rides through them, into a district where the butterfly fauna is impoverished. It had been the little-grazed Inclosures which had carried such magnificent butterflies and on which the halcyon generations of collectors had descended between 1870 and 1960. The number of ponies in the Forest was big enough in the early 1950s for them to start clearing the brambles, reducing the blossom previously alive with insects.[xxiii] The Commission was already clearing undergrowth even before the ponies were let in and, whereas livestock found wandering in the Inclosures were impounded until the late 1950s, they were tacitly admitted from the mid-1960s and especially after 1969. The fences around half-a-dozen large Inclosures were removed between 1970 and 1974. Substantial butterfly populations may have persisted a fraction longer in the New Forest than in much of the countryside outside, but what soon happened was of a piece with the intensified tidying of the entire landscape in the decades after the Second World War.

Although it is true that trees would eventually have grown up in the Inclosures, creating more shade than in the late Victorian heyday, the Forestry Commission would have been able, had it wished, to plant and fell so as to maintain a flow of suitable habitat. From this point of view, it was unfortunate that the Commission came into the hands of men whose exclusive interest was timber production. In Wytham Woods, the Oxford University ecologists won a long battle against the University's own foresters, but the New Forest had no equivalent champions and experienced no such victory. The Forestry Commission and the National Trust are said to have been energetically and expensively destroying ancient woodland in order to replace it with conifers. In 1967, the Commission bombarded a Suffolk wood with Agent Orange to try to rid it of ancient tree stools thought to compete with conifers.[xxiv]

In England as a whole, butterfly fortunes were influenced by the virtual cessation of coppicing. The views of C. J. M. Bowen about such matters are especially interesting. As a chemist, Bowen was sceptical about the effects of chemical sprays.[xxv] He noted that Bernwood Forest, Oxfordshire, had been subjected to intensive aerial spraying by the Forestry Commission yet remained one of the richest entomological sites in the South Midlands. He realised that, in Britain, most butterfly species are near the northern edge of their range and exhibit wide fluctuations in numbers from year to year, which puts them at risk if barely discernible negative factors coincide. They are susceptible to small alterations in temperature, besides unobserved influences, like incidences of parasitism. In such circumstances, it is dangerously easy to think that any decline must be permanent or attribute it to some single conspicuous event.

Bowen thought that even the effects of destroying habitats were speculative. Referring to woodland species, he claimed that the availability of food plants does not seem to be much of a limiting factor and noted that the area of woodland in the southern counties has remained constant for centuries at about 10% of the land area. Declines might be explained by changes in management such as the reduction of coppicing, the replacement of deciduous trees by conifers, and the rearing of pheasants (which continues to expand). But deprecating these phenomena, as is usual among entomologists, did not square easily with the recent increase (Bowen was writing in 1979) of four or five woodland species.

Big swings in the fortunes of farming, often mediated for grassland species through what happened to rabbit populations or transmitted as a secondary effect to woodland species by the falling demand for sheep hurdles, do seem to be implicated in losses. The effects show up in local studies that track alterations in habitats over several years. Once again it must be stressed that different species are differently affected, but the conclusion is that if there ever was a Golden Age for butterflies this came during the long arable depression from about 1870 to 1940. It is of course possible to argue that the gains of that period, as the naturalist conceived them, were an unintended consequence of importing food from the Americas and Australasia, and that the other face of this was exporting the environmental costs. The prairie, outback and pampas bore the costs, while English naturalists reaped the landscape benefits of agricultural decay. We shall turn in later chapters to Europe's expansion overseas and the environmental history of North America.

Against the long depression, the present situation in England cannot compare. Even allowing for the exaggeration of modern losses, butterfly enthusiasts and other naturalists seldom experience anything to compare with the profusion of those years. But if we look before the long depression and consider the wood-pasture regime of early modern times, the current period may not seem quite so dire. The conventional view is that current farming practice involves fierce technologies, like the flailing of hedges and consequent destruction of butterfly larvae, by contrast with which ancient technologies were feeble. Agreed, but this does not allow for the ubiquitous distribution of the human population when England was a genuinely agricultural country and the use it made of any and all natural resources. That the present status of butterflies is the latest stage of an uninterrupted contraction of range and number is highly unlikely.

Historical fluctuations hold little interest for environmentalists, who fear that the twenty-first century is the time of a great and maybe the final extinction of life. Estimates of the rate of extinction have however repeatedly been challenged, although unavailingly — it is scarcely disinterested science that produces them in the first place. Roger Sedjo, a forestry economist with Resources for the Future, showed that alleged deforestation rates in the rain forest are inflated more than two-fold and the number of species lost is not directly proportional to the area of forest cleared.[xxvi]

None of this carries weight with environmental lobbies. I can do no better than quote a letter by A. J. Perry in the *Financial Times* of 12 April, 2007. He points out that the natural world is outstandingly adaptable and we try to offset changes at our peril, while the accuracy of doomster predictions has been abysmal. Of the threatened effects of global warming, Mr Perry asks, 'would someone please speculate about the wonderful new flora and wildlife that would flood into this new watery, muddy environment, as well as the economic benefits that might arise?'

ENDNOTES

[i] M. A. Shannon, *The Aurelian Legacy: British Butterflies and their Collectors* (Berkeley: University of California Press, 2000), pp. 17–18.

[ii] P. S. Savill *et al.*, *Wytham Woods: Oxford's Ecological Laboratory* (Oxford: Oxford University Press, 2010), p. 72.

[iii] Shannon, *Aurelian Legacy*, p. 18.

[iv] Quoted by M. Oates, 'The Demise of Butterflies in the New Forest,' *British Wildlife*, **7** (4), 1996, p. 208.

[v] D. Tomlinson, *Britain's Butterflies* (Old Basing: Wildguides, 2002), p. 26.

[vi] UK Butterfly Monitoring Scheme (UKBMS) 2011. Survey of changes table available online. The sample size may however disguise declines among very localised species.

[vii] E. B. Ford, *Butterflies* (London: Collins New Naturalist, 1945), pp. 138, 140.

[viii] I am the first to accept that too many developments are poorly located. A single outrageous example demonstrates the point — that sanctioned close to the Richard Jefferies Museum at Coate, Wiltshire.

[ix] J. Thomas and R. Lewington, *The Butterflies of Britain and Ireland* (Gillingham, Dorset: British Wildlife Publishing, revised edition 2010), p. 63.

[x] E. Anderson, *Plants, Man and Life* (Berkeley: University of California Press, 1952).

[xi] J. Asher *et al.*, *The Millennium Atlas of Butterflies in Britain and Ireland* (Oxford: Oxford University Press, 2001).

[xii] Ford, *Butterflies*, p. 133.

[xiii] E. Pollard and D. Moss, 'Historical records of the occurrence of butterflies in Britain: Examples showing associations between annual number of records and weather,' *Global Change Biology*, **1** (2) (1995), pp. 107–113.

[xiv] J. A. Downes, 'The History of the Speckled Wood Butterfly (Parage aegeria) in Scotland, with a discussion of the recent changes of range of other British butterflies,' *Journal of Animal Ecology*, **17** (2) (1948), pp. 131–138 (Abstract). I am indebted to Dr Jane Hill for this reference.

[xv] Asher, *Millennium Atlas*, p. 346.

[xvi] Tomlinson, *Britain's Butterflies*, p. 96. Concerns about the Muntjac do not seem to have been realised.

[xvii] L. Hugh Newman, *Living with Butterflies* (London: John Baker, 1967), p. 203.

[xviii] J. P. Dempster, 'The scientific basis of practical conservation: Factors limiting the persistence of populations and communities of animals and plants,' *Proceedings of the Royal Society of London, Series B*, **197** (1977), pp. 69–76, Abstract.

[xix] Thomas and Lewington, *Butterflies*, pp. 198–199.

[xx] L. G. Higgins and N. D. Riley, *A Field Guide to the Butterflies of Britain and Europe* (London: Collins, 2nd edition 1973 [first edition 1970]), p. 21.

[xxi] R. South, *The Butterflies of the British Isles* (London: Frederick Warne, 1941 edition); Ford, Butterflies; E. B. Ford, 'Woodland Butterflies,' *The New Naturalist: A Journal of British Natural History* (London: Collins, 1948), pp. 47–50; Asher, *Millennium Atlas*; Tomlinson, *Britain's Butterflies*; and Thomas and Lewington, *Butterflies*.

[xxii] Oates, 'Demise,' p. 215.

[xxiii] C. R. Tubbs, *The New Forest* (London: Collins: New Naturalist, 1986), pp. 188–197, p. 270; *The New Forest: History, Ecology and Conservation* (Lyndhurst: New Forest Ninth Centenary Trust, 2001), pp. 224–232. The latter version was written in 1997.

[xxiv] *Financial Times* 21 July 2012.

[xxv] C. J. M. Bowen, 'Butterflies of Berkshire, Buckinghamshire and Oxfordshire,' *Transactions of the Newbury District Field Club*, **12** (5) (1979), pp. 61–62.

[xxvi] L. Kaufman and K. Mallory (eds.), *The Last Extinction* (Cambridge, MA: The MIT Press, 1986), p. 12.

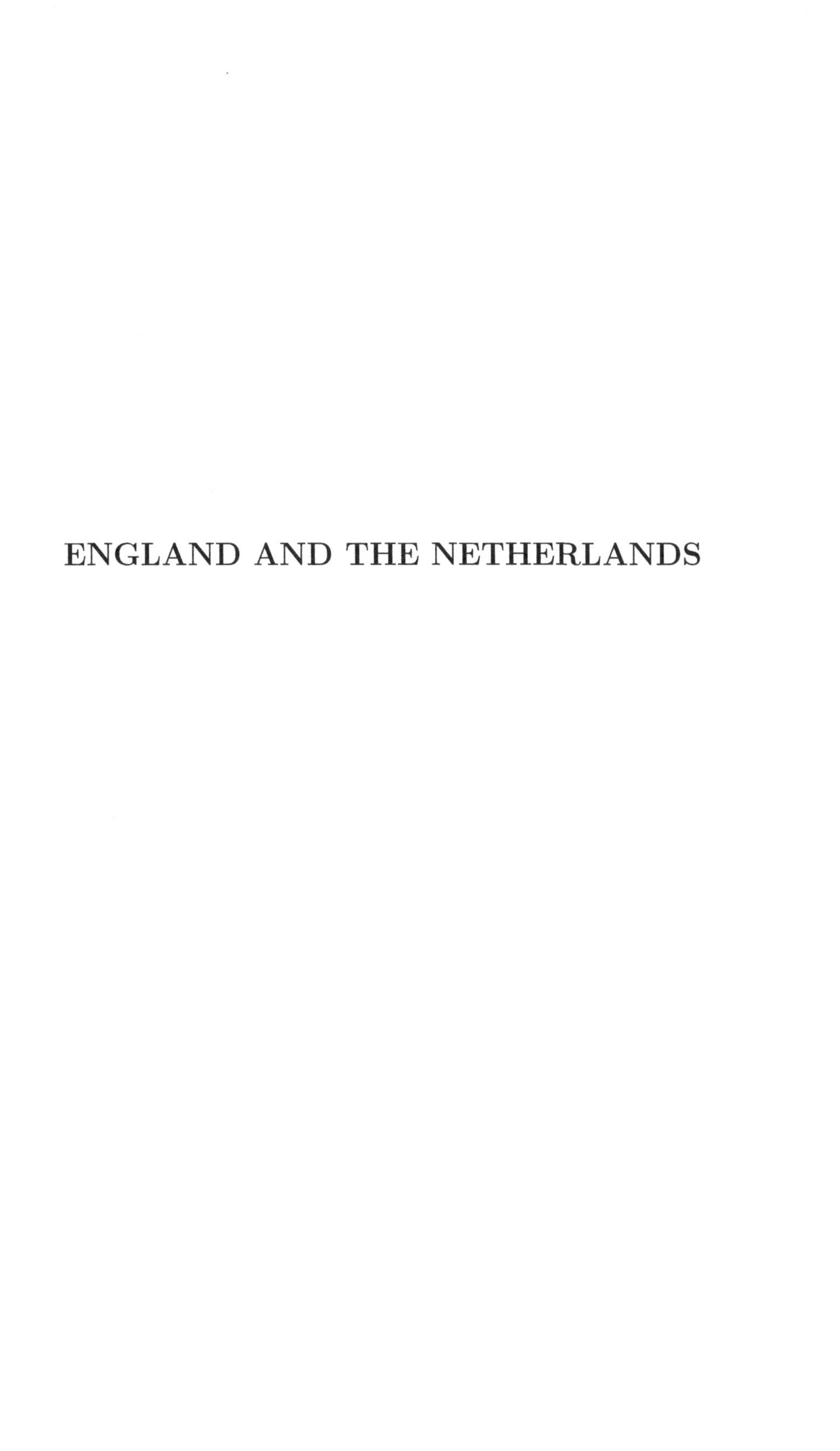

ENGLAND AND THE NETHERLANDS

4. COMMODITY LANDSCAPES: SOUTHERN ENGLAND

Given that [Lapwing and Skylark] benefited from the felling of millions of acres of Europe's wildwood, the birds' airborne displays were, in a sense, a hymn to Neolithic man's agricultural triumph.

Birds Britannica (2005)

The level topography, tillable soils and reasonable climate of the hinterlands of Amsterdam, Paris and London were undoubtedly superior to areas north and west, yet the influence of these endowments is hard to separate from the effects of urban demand or new husbandry techniques. Intensive methods such as row cultivation and the sowing of various fodder crops spread during the middle ages. In addition, modes of landownership and the organisation of rural society modified the productive character of the soil, distorting and even dominating its raw potential.

Innovation was slow. Until the nineteenth century, there were few institutions devoted to pooling and sharing information about best practices. Large farmers and landowners across Europe did, however, buy the books that had begun to appear during the seventeenth century, and their marginal comments show they took the authors' ideas seriously.[i] Lesser farmers, the great majority, were not literate or rich enough to do this for themselves, though they may have followed their landowners' leads or bowed to their exhortations. Otherwise, much of the adoption of new crops and bewildering range of new rotations may have come about through 'stimulus diffusion'. This amounts to trialling ideas that have been seen over a neighbour's hedge or spoken of as worthwhile; it is half way between original invention and the importation of novelties from elsewhere. Direct importation was most apparent when the Low Countries adopted the methods of Lombardy. Eastern and Southern England brought them on from Holland.[ii] The sequence is familiar in outline, though once more the respective contributions of farm size, layout and tenure, as opposed to agronomy, livestock breeding and methods of tillage, elude universally accepted generalisations.

All the aspects varied kaleidoscopically from time to time and place to place.

English agriculture became decidedly more intensive from the sixteenth century. Whereas in 1500, little fodder had been grown and livestock were often scavengers, pasture management was improved (especially after about 1650) by sowing a range of legumes often grouped under the umbrella term, clover.[iii] Fodder plants had previously been located according to their suitability for different soils. This equilibrium was disturbed by the improvements in herbage, which permitted redistributions of agricultural ecology. The wider botanical implications and effects on the organisms formerly living in tracts of rough herbage remain obscure.

Consistent with agricultural intensification was a slow dwindling of concern with the intermediate goods that medieval landowners had been eager to produce close to home, such as pigeons harvested in dovecotes for their eggs and squabs (young), or carp reared in fish ponds.[iv] Hunting and shooting were at first of practical significance. 'The importance of the produce of the field sports of antiquity can hardly be over-emphasised, as far as the rich — and only the rich — were concerned', wrote the twentieth-century Warwickshire landowner, Michael Warriner.[v] Venison, pheasant, partridge and duck all supplied food quickly in the winter, whereas the carcases of livestock, hung for longer, could more easily go bad. (It was impolite to comment on the mustard used to disguise the taste).

Later the landed class ceased to rely on the produce of hunting on their estates, or from fishponds, or dovecotes. Ice-houses meant meat could be stored much longer, and agriculture was becoming more productive. During the eighteenth century, orchards and warrens were gradually moved away from country houses and out of their parks. Fish ponds were less movable but those in sight of the big house could be turned into ornamental duck ponds. An Act of 1765 'for the preservation of fish in fish ponds and other waters and conies in warrens' was seemingly ineffective. Types of food formerly acceptable, like rabbit, fell in status to become food for the poor. The alterations were connected with the better farming, more extra-agricultural sources of income, and the growth of trade (for instance sea-fish became widely available). Yet, as they ceased to rely exclusively on local sources of fresh meat, so landowners began to rear huge pens of pheasants for the sake of the shooting battue. Access to the land was increasingly denied to the poor, this time in the sacred name of sport.

COMMODITY LANDSCAPES

One way of looking at the relationship between people and wildlife is to think of the landscape at any given date as if it were devoted to a single commodity, changing as markets changed. This may strike people whose first interest is the natural environment as overgeneralised and abstract, but it has the advantage of clarifying why landscapes and ecosystems altered from time to time as well as differing from place to place. The humanised landscape which now occupies most of the earth's surface cannot be understood solely in terms of the dictates of nature.

The commodity concept is grounded in the natural endowment but the mediation of relative costs is needed to explain why cultivated landscapes vary so much. For example, different soils and climates — meaning different costs for farmers — may favour plantation crops or grain or livestock.[vi] The industries associated with the respective types of production will vary, as will factors of production, location of settlements, distribution of income and so forth, up to and including the entire trajectory of economic development. Although the notion is unavoidably mechanistic, and for completeness would need to take each historical period into account, these economic patterns require only a further step to draw a picture of the general variety of ecosystems.

The idea is akin to staple theory, in which the mode of production is determined by the dominant crop: the Canadian sequence (which was the type specimen) ran through fishing, fur-trapping, lumbering and grain-growing, the Australian sequence through marine products, wool, gold and grain. Each product required different inputs, was harvested from settlements located in different ways and prospered according to the world market for the commodity in question. The scheme was originally devised by the Canadian economic historian, Harold Innis. We need not follow the fate of staple theory at the hands of its critics to recognise a potential implication: the landscape characterised by a given commodity houses a specific ecology. When markets favour a fresh crop, the 'living entourage' is transformed as well. Certain species find congenial niches and prosper with the new commodity, but others fail to do so; their numbers fall and their distribution shrinks. The idea is admittedly incomplete, thanks to unexpected limits of plasticity or adaptability on the part of some wild species, and to the fact that cost schedules must in part reflect underlying nature. At its extreme, the staples idea might be dismissed as a deterministic thesis in which ecosystems are solely functions of economics; even so it remains suggestive.

A sequence of staples describes the pattern of export production in new lands better than it accounts for farming in the ancient entanglements of long-settled Europe. If we think back to sixteenth-century England, we find a mixture of products for household consumption or very local sale and a proportionately small volume of goods despatched to customers elsewhere. Rural households made use of every growing and living thing; because they were poor they could not afford to miss any opportunity. They could afford to harvest local nature because the opportunity costs of female and child labour were low — there was little else for them to do on the spot and they could not travel far to work. Local subsistence production continued quite late in some districts yet was not wholly distinct from production for distant markets. The pattern was seldom clear-cut since product markets and long-distance trade had existed far back in prehistoric times. Supplying the markets meant that some specialised and large-scale businesses arose to interact with the ecosystem and harvest its gifts.

Southern England is suitable for analysis because wide areas within it, notably the Cotswolds and Wessex chalk, experienced swings in land use resulting from alterations in relative product prices. From perhaps as early as the end of Roman times, these gentle uplands, settled and cropped during the Iron Age, turned into gigantic sheep-runs, whereas from the seventeenth century they were once more increasingly ploughed. Cultivation was at its historical height in the 1850s and 1860s, when cereal prices were high. J. G. Cornish, rector of East Lockinge, Berkshire, harked back to his boyhood home in Suffolk of which he said, 'the intensity of cultivation undoubtedly diminished the picturesqueness of the country. One hardly ever saw an acre of land that was not cultivated'.[vii] His observations fit any arable district. Such was the land hunger that the big straggling hedges were grubbed up, and the wide ditches filled in, to be replaced by straight and narrow gullies. Miles of wide grass verges beside the country roads were annexed by the adjacent landowners and quickset hedges were planted beside the now pinched roadways. Hedgerow trees were felled 'ruthlessly'. The term is Cornish's and he knew of a whole attractive wood grubbed up for corn land. It was the cereal grower's Golden Age. When he came to know the downs on the Berkshire chalk, Cornish said, 'it was like the good times in Suffolk on a more lavish scale.'

Successive depressions during the nineteenth and twentieth centuries reversed the situation with a vengeance, creating large areas of tumbledown arable. Ornithological records indicate that these periods harboured birdlife markedly different from the phase of cereal prosperity. In 1881, a couple of years after the onset of deep depression, Richard Jefferies saw that superficially there was little change, only a few spots having gone right out of cultivation. Most land continued to be tilled, but not so well, and while there were still plenty of cattle, they were fewer than before and of poorer quality. Ominously, however, the old tenant farmers had often quit and been replaced by less experienced men. From then on, agriculture on the chalklands went from bad to worse. Sheep runs replaced the highest upland fields that had been brought under the plough in the mid-nineteenth century. Crops on the remaining cultivated acres were no longer carefully weeded. The region became the quiet countryside of reminiscence that fills so many rural books, where the cereals were always bright with poppies and full of corn cockle. Big flocks of linnets liked the weed seeds and so did pheasants — it was the 'Age of the Pheasant', maximum game-keeping, and huge bags in the shoots.

The clay vales that lie between the chalk and limestone are harder to study. The long-term history of Whiteparish, Wiltshire, has been the subject of a detailed investigation which is indicative. It shows that the less pronounced land-use changes in the vales seldom offer the means for historical 'experiments' like those of the uplands.[viii] Two-thirds of Whiteparish have a drift (surface) geology of Reading Beds, London Clay and Bagshot sands, all of which had been heavily forested in early times. In the Saxon and medieval periods, parts of the forest were slowly cleared. During the seventeenth century, fields were enclosed in the remaining wooded areas and on the downland, and new farms were established. The nineteenth century saw the enlargement of one park in Whiteparish and the making of a second; otherwise most eighteenth- and nineteenth-century developments related to the enclosure of existing farmland, including the amalgamating of small holdings that had been allocated at the Act of enclosure in 1804. This created the large fields (now even larger) which characterise the Whiteparish landscape. Overall, the experience of this parish, at least away from its downland end, was of subdued land-use changes unlikely to produce effects that will show up dramatically in the biological record.

While land-use fluctuations in the one-third of the parish that lies on the chalk may have approximated the see-saw experience of most southern English downland, what is noticeable about the remainder of Whiteparish is its incremental development. It is a model for the clays of southern England in general: woodland was slowly cleared, farms and fields were regularised and landownership became less equal. The Dissolution had brought monastic properties into the hands of speculators who went in for dis-parking on a sizeable scale, but the work needed to convert parkland into productive farms, although undertaken by successive lessees, took generations to complete.[ix] Big shooting estates on the uplands are more likely to generate biological records — often records of birds of prey killed because they were supposedly dangerous to game-birds. When changes in land use occurred on the downs, they tended to be relatively rapid. The vales are harder nuts to crack if we wish to investigate, or merely to sketch, the natural history impact of land management.

One study did however make the attempt and astutely utilised the Whiteparish example.[x] It was an investigation of populations of banded snails, *Cepaea,* which was able to demonstrate their selection for camouflage in stable habitats. *Cepaea* are favourite organisms of evolutionary geneticists because of their considerable colour variation. Continuity in the snail populations of the clays permitted the expression of variation, which unstable habitats on the down land showed less clearly. Modern *Cepaea* populations were found to differ according to landscape history. Snail populations are long-lasting in the ground cover. For larger, mobile categories of wildlife, deductions about the past would be more problematic.

Once the Romans left, human inhabitants deserted the high ground on the chalk and limestone and occupied villages in the valleys. Saxon villagers continued to grow their own cereals but some entrepreneurs took to grazing flocks on the wolds. In the Middle Ages, they were visited by purchasing agents from as far away as Flanders and Italy, seeking to buy their wool clip. The great sheep farmers and local merchants were the ones who endowed the 'wool churches' of the Cotswolds and whose flocks are most visible in the historical record, though the aggregate number of sheep owned by the peasantry was much greater. Sheep, W. G. Hoskins tells us, were to be found everywhere but concentrations occurred on the chalk and limestone, on other areas of light soil (such as the Brecks) and on salt marshes.[xi] The medieval desertion or clearance of villages in the Midlands in favour of wool production resulted in the abandonment of one-tenth or even one-sixth of them and radically altered the landscape. Among the wool districts, the Cotswolds figures most prominently in the record because the quality of its wool attracted merchants from European manufacturing areas. The classic account of merchants both local and foreign is Eileen Power, *Medieval People.*[xii]

As its county historian, John Steane, remarks, the large-scale turn from arable farming to sheep grazing in Oxfordshire during the late middle ages markedly reshaped the landscape.[xiii] He adds a few details about monastic estates. Osney kept a total of 3,000 sheep in Oxfordshire and Gloucestershire, its various estates coming to specialise in different aspects of the production process. The wool clip was gathered at Water Eaton, sent overland to Henley, and down the Thames to Italian merchants in London. Medieval trade routes already tracked north-west to south-east, funnelling down to the head of Thames at Lechlade, along such routes as Tame's way, which ran across from Barnsley to Fairford and Lechlade and bore the name of the rich wool entrepreneur, John Tame. He was the late fifteenth-century merchant who paid for the stained glass windows in Fairford church, the best medieval glass in the country. Welsh sheep and ponies followed the same route, which led ultimately to the London market.

Early in the sixteenth century, textile production began to be concentrated in great patches within England and sheep rearing remained of the first importance. What precise changes in animal and plant life may have resulted we can only speculate but that they were significant seems undoubted. Two or three seventeenth-century accounts of the close-cropped downland, especially that by the Wiltshire antiquary, John Aubrey, signal a complex of wildlife very different from that on tilled fields.

When the economy picked up after the Civil Wars of the mid-seventeenth century, as it did quite rapidly, a change in farming began on the Cotswolds and Wessex chalk. What happened thereafter was the progressive ploughing of down and wold pasture and its assimilation to a mixed farming regime. Less attention was paid to wool from grass-fed sheep; multiple goods were produced ever more intensively from what was turning into a mixed commodity landscape. The emphasis was less single-mindedly on wool and more on meat and manure, produced partly on the plough land and partly on the down grass. The archetypal arrangement was for sheep to be fed on the upland grass during the day and folded at night (penned in hurdles) on the arable land, where they consumed fodder crops and dunged the soil. This raised soil fertility for subsequent cropping, largely by transferring it from the downs, with long-term effects on the distribution of soil fertility that are highly credible but beyond our ability to document in detail. It may be significant that farms out on the downs are not infrequently called Starveall or some similar name, while fertility seems to be greatest around the older village farms.

Wool, mutton and manure were joint products and only occasionally is it possible to tell with certainty which was the chief aim. Maybe there was no dominant product since fluctuations in costs and relative prices endlessly recalibrated the farmer's purposes and there was security to be had in reducing dependence on a single product. Peasants had long engaged in the practice of folding, i.e., penning, their sheep on the plough-land overnight to fertilise the soil, having run them to graze down-land grass during the day. From the seventeenth century, growing new fodder crops (clover and other species) on the arable land meant that the sheep numbers involved could rise. In autumn, the flocks were driven across the Wessex downs to huge fairs like Weyhill, Hampshire, and East Ilsley, Berkshire, where they were bought and moved closer to London for fattening. Selling the sheep off the farm may on balance have removed as much fertility as their dung provided but the crops, trodden in, did build up soil nitrogen.

Permanent grass was no longer so vital for feeding purposes. Arable farms began to be built in remote situations on the downs, well away from the villages. Horses could be stabled on the spot. This avoided daily journeys from the village farmsteads, which burned up energy in feeding the teams of horses. Paradoxical as it may sound, the upland grass of lowland England was the subject of a form of internal colonisation, the highest and thinnest soils finally succumbing to the plough during the Napoleonic Wars, with new-built farms dating from that period. The formerly open grassland of the Cotswolds began to be confined by stone walls, because (except during the war years) labour and stone were both cheap. On the chalk, where there was no building stone, gappy hawthorn hedgerows had to suffice. It was a shift from open prairie to a landscape of what, by English standards, are immense enclosed fields.

Away from the remoter wolds and downs, the growing of crops became more varied and intensive. In the Vale of White Horse, Berkshire, livestock products were emphasised but there was also plenty of cereals from the arable land round about. Pastoral parts of the Vale landscape supported the Wantage tanning industry, which housed one of the largest tanneries in the country. Until the owner went broke in 1811, Paul Sylvester (with the help of a German named Desmond) halved the time it took to ready hides for the making of leather goods. Meanwhile dairying encouraged the production of pigs. They were fed on dairy by-products. Many were slaughtered each year in Wantage and Wallingford, though the greatest cluster of bacon houses was Edward Loveden's in Faringdon, which cured over 8,000 sides of bacon annually.

The problem for milk producers was that liquid milk would not keep for long and before the railways could not be carried far to market, hence the diverse ways of converting it into more lasting forms. One was to make cheese: the making of Double Gloucester spread eastwards into the Vale from its home county. Vale farms added cheese rooms with large numbers of cheeses stored to mature on racks. Closer to London, other distinctive landscapes appeared, such as stretches of market garden and pastures outside the city in Middlesex. Farm leases mandated sowing grass on land from which gravel had been extracted for the building trades.

Barley growing in and around the Vale of White Horse supported malthouses in the towns. When the malt was not sold to the large number of local brewers, it was shipped down the Thames to London. A prominent commodity of the late seventeenth and eighteenth centuries thus created an apparatus of processing plants. Town halls were built in many southern towns in the years between 1650 and 1670, which marked the recovery in the local economy after the Civil War, the growth of trade on the Thames and the rise of intensive mixed farming. At Abingdon, partitions for further corn dealing were made in the market hall in 1729. The Thames bank at Abingdon was lined with wharves and warehouses, as was the bank at Lechlade, the head of Thames navigation. All this was on a miniature scale by modern standards, though the total effect in the arc around the metropolis was not to be sneezed at. Fortunes were made in the process — the Tomkins family, Baptists at Abingdon, certainly made theirs out of malting.[xiv] Most towns had some malt-houses; Wallingford had 18, besides one of southern England's major breweries in Edward Wells' establishment.[xv]

We are starting to see in the range of processing industries in the towns a reflection of the mosaic of farm enterprises, still capable of differentiating landscapes but with the facets no longer as monocultural as when sheep kept for their wool had virtually monopolised the downs and wolds. Mixed farming offered joint products and, as already mentioned, sheep were increasingly fed off fodder crops like clover and turnips rotated with cereals on the same farms. The advantages were that the by-products of each enterprise — crops, sheep dung — could be consumed within one and the same farm. This evolved into the standard mixed system of Victorian agriculture. The enterprises involved were highly commercial, aimed at supplying the London market, shifting with its demands and capable of 'ruthlessly' stripping nature out of the countryside if a little more cropland promised to make the farmer rich.

Over the centuries, habitats were reshaped by changing prices. The same districts grew certain commodities in one period and others when demand shifted. They were intensively farmed when prices and expected profits warranted it and left to rack and ruin when arable farming ceased to pay, for instance in the depression of the late nineteenth century. At that time, we start to find accounts of wildlife adapting to semi-neglected or vacated farms. If there was an alternative, as there was on the Vale clays once railway transportation made it possible to ship liquid milk to London, it was to convert the fields and equip the farms for dairying. Conversion to more specialised dairying took place along the Upper Thames around Cricklade, Wiltshire, as may be traced in sale catalogues of the time. Richard Jefferies noted of this vale country that the rough, wet patches, full of rushes, thistles and grasses, were being grubbed up and drained to grow hay. Wildlife and wild flowers suffered in the vales just when on the downs falling cereal prices were rescuing them from the increased ploughing described by J. G. Cornish.[xvi]

WILDLIFE HISTORIES

Describing the chalk and limestone as commodity landscapes is most appropriate at the height of the medieval wool trade but is intended to be a general organising framework, a means of rejecting any notion of untrammelled nature and conveying the idea that ecosystems were shaped and re-shaped by markets for different types of produce. It is not intended to suggest dogmatically that the land had no uses beyond grazing sheep or growing local food. Another long-standing use was for rabbit warrens.[xvii] These had tended to appear on poor land in remote places, on grass slopes too steep and thin-soiled to cultivate — except during the Napoleonic wars, when warrens were ploughed up in favour of crops of grain. So-called pillow mounds had been built, with artificial burrows to accommodate breeding does. Those that survive are marked on the Ordnance Survey maps; on the ground they look like elongated air raid shelters left over from the Second World War. Hay was brought to them in severe weather. The warrens were commercial operations and had grown in number during the fourteenth and fifteenth centuries, when the human population had decreased after the Black Death, when marginal land had fallen out of production and landowners were pleased to be able to diversify the activities on their estates.

Monographs on the rabbit concentrate on biology and where they touch on the relationship with man tend to do so through accounts of warrens, pest control, land law and poaching. What is missing is a focus on markets and consumers, which would explain why rabbits were farmed in the first place. Writers refer to farming the animal in warrens but soon return to discuss its biology. It is therefore not always made clear how far the species is an anthropogenic one, its status, distribution, even its very presence in the British Isles, being dependent on human activity. Rabbits had probably first been introduced from France in the late eleventh century. The fullest historical coverage is by Sheail, who has pieced together innumerable references from scattered and obscure sources.[xviii] Anyone who seeks information on ecological history is obliged to employ similarly arduous methods; fortunately in the case of the rabbit we do have Sheail's work as a basis.

The post-medieval history of the rabbit population was, for a long time, largely an account of their artificial preservation in the warrens. Since dry grasslands were so suitable for sheep, the rabbit population did not make enormous headway and was seemingly not large — according to Gregory King's possibly fanciful figures it was in total only one-eleventh as large as the national sheep flock in the 1690s. Sheail has used the Board of Agriculture reports for each county to compile a map of the distribution of warrens in the 1790s. The map is incomplete (for instance, Wiltshire is a blank) but a correspondence with areas of remote or poor soil is evident. By that period, however, even thin soils were being brought under cultivation and the number of warrens and total population of the rabbit was shrinking. It recovered, nevertheless, because rabbits could feed on the rich fodder crops sown as parts of arable rotations, while harbouring in remaining unploughed areas and along the hedgerows of enclosed fields. Fodder crops filled part of the 'hungry gap' in spring, before the new season's cereal crops had grown. They were grown to feed farm stock but rabbits were unintended beneficiaries.

Escapes from the warrens, or from the mass slaughter when warrens were being destroyed, scattered rabbits into the countryside. Sheail supposes that, by the 1850s, the wild population had become larger than the warren population of a century earlier. Yet rabbit demography was never linear. In prosperous times, it paid the farmer to have men control the rabbits, since they took a heavy toll on crops, but he could not afford this during the depressions of the nineteenth and twentieth centuries, and at such times, rabbit numbers quickly expanded. Left to their own devices, the population rose until the wild burrows became densely populated and the numbers were pruned by savage epidemics. Trappers made a living by catching rabbits for sale and their take may have dampened these cycles. The Ground Game Act of 1880 permitted farmers to take rabbits and the invention of the gin trap made catching them easier. Nevertheless, the rabbit had the upper hand and had spread throughout Britain by the mid-twentieth century. It was a phenomenally abundant species that may have been at its peak in 1953. Then farmers imported and spread the virulent disease of myxomatosis, which filled the countryside with blinded and dying rabbits and caused the population to crash.

Until that time, the rabbit had been a significant element in rural ecology and economy. In 1953, I crested a rise in the downs of north-west Hampshire and saw the entire hillside move like a grey tide. The thin, flinty, chalk soil was covered with rabbits as numerous as those shown in a famous photograph of them encircling a dam in Australia. Gin-traps were set on the downs but could take only a tithe of the population. They caught other birds and animals too and I found a green woodpecker in one. By contrast, myxomatosis was total war, like bombing the countryside with napalm, and just as distasteful. Alternatively, it was a neutron bomb that left the land but eliminated the animal inhabitants. When the rabbits were almost gone, as they were in two seasons of crouching, eye-bulging, blinded victims everywhere one went, a complete ecological ensemble on the surviving fragments of downland went with them: the orchids were choked by the growth of coarse grasses and the birds of heavily grazed grassland, wheatears, whinchats, woodlarks and wrynecks, melted away.

The wryneck had been in long-term decline since the 1870s; it was not one of the beneficiaries of depressed agriculture. The tall grass of neglected pasture overshadowed the ants' nests on which it fed and drastically reduced their size. By the 1950s, wrynecks were rare and I spent several evenings admiring one when I should have been revising for A-level examinations. The whinchat loved the neglected pastures, dotted with small hawthorns, of the 1920s and 1930s but these were ploughed out and, when myxomatosis had struck, alternative habitats became overgrown with rank herbage.[xix] I could rely on finding whinchats on the Lambourn Downs and woodlarks on the Hampshire chalk until the 1960s but not afterwards. The wheatear, too, was badly hit by the wartime and post-war ploughing that reduced the area of down grass.

In the 1950s and 1960s, farmers told me how much they itched to cultivate every possible inch; ploughing grants encouraged them. Some did not scruple to carve into the environmentally rich scraps of land that, in the brave dawn of the Nature Conservancy after the war, had been designated (but not protected) as SSSIs (Sites of Special Scientific Interest). An eminent member of the Conservancy was later to rue that the threat to the environment from the intensive, uniform, herbicide-wielding agriculture of the post-war years was simply not anticipated.[xx] But on the ground the threat was surely obvious; but agricultural protection perpetuated the damage that humbler employees of the Conservancy could see full well.

Rabbits gave the joint products of fur and meat, the markets for which did not always move in phase. Rabbit fur was suitable as a covering for hats and caps and was cheaper than importing furs from the Russian woods or beaver from North America. Rabbit meat was a valuable item of diet, not just for the rural families who could not have got through the interwar depression without it, but in trade to the cities. There is a temptation to think that the meat could not be transported before the railways, or refrigeration, but in the seventeenth century, rabbits were being carried to London from a wide swathe of the country. Letters for London from the Isle of Wight were entrusted to the 'coney-man', who crossed the Solent once or twice per month with rabbits to sell.[xxi] The meat markets of northern towns and cities were provisioned from the Yorkshire Wolds.

By the 1820s, the trade was highly organised and middlemen had farmers or warreners under contract: the countryside was no backwater exempt from the marketing networks of urban society. Furs were despatched to the furriers of Stamford Bridge and Malton. Elsewhere in the country, there were towns specialising in processing fur, and in the 1860s, there were 80 fur-cutting machines operating in London, where the trade employed 8,000 hands. A Select Committee of the House of Commons was informed in 1873 that, every year between October and March, rabbits totalling 2,300,000 were supplied to eight urban areas in the Midlands and north of England.

England produced sufficient silver-grey pelts, besides the common grey rabbit, for them to be sent in bulk to northern Europe and the Far East, though this trade declined by the late nineteenth century. By that stage, English rabbit production could no longer meet home demand, and skins were being imported from Germany and other parts of northern and central Europe. Belgium had been exporting to England from the 1840s, and sizeable supplies came from Ireland and Scotland by the mid-1850s. This was a tribute to the growing purchasing power of London, which was the great entrepot around which world trade in rabbits revolved. Several other biological commodities experienced a similar turn-round in which English demand could no longer be met by domestic supply and this had to be supplemented from continental sources. Nor was it merely Europe that sent supplies, by 1880, skins were being received from Australia and New Zealand.

It took only the invention of refrigeration for the stream of imports to become a torrent, no longer merely supplementing but under-cutting English suppliers. In 1906, Australia was earning more by exporting rabbit meat and furs than by shipping beef.[xxii] During the 1920s, an average of over 9 million rabbits were shipped out annually, a trade that was to become worth 3 million pounds per year.[xxiii] Rabbit damage in Australia, though, was guessed to cost 10 times more. In England, meat and fur prices fell, though not so far as to destroy the local trade. Rabbits were trapped or shot, and until the 1950s, carcasses could be sent by train or through the post with nothing more than a label attached to their hind legs. Long afterwards, I talked to a woman in her family's café in Warminster, premises which they had used until myxomatosis for a rabbit-packing business. My sympathy about the loss of the trade was misplaced; she had hated handling the rabbits and said the 'myxy' epidemic was the best thing that could have happened. Thereafter, the public was wary of eating rabbits, but until that time, hers had been a respectable trade. The margins and slopes of arable areas had formed a fringe commodity landscape for rabbit meat and fur.

VERMIN

Rabbit warrens never occupied a high proportion of the land surface but they were long a characteristic feature of the scene and a target for vermin, wild and human. The concentrations of production meant that fortified warreners' lodges were built in places to protect against gangs hoping to steal the stock. Information about this history is ample but scattered. Other species of pest offer better collections of data on the wildlife of early modern England.[xxiv] These data are contained in the vermin accounts. From Tudor times, parishes were empowered to make payments for the tails or corpses of species adjudged to be vermin or pests — one has to say adjudged because some entirely harmless animals were included. Details survive among parish papers, sometimes running into thousands of individual payments in the case of house sparrows, but applying to a wide range of birds and mammals (such as badgers and foxes) that were thought to be damaging to agriculture or horticulture. It was a means familiar in agriculture whereby costs were passed on from farmers to local taxpayers. Farmers were as vociferous about supposed sources of loss as they have become about culling badgers. But vermin payments petered out in the 1830s, by which time the cost no longer struck people as worthwhile.

Writing at Alveston in the Severn Vale in 1829, John Knapp was perfectly aware of the nuisance caused by sparrows but thought they were easily dealt with and considered that paying to destroy other species had become redundant.[xxv] Whereas it had once been vital to keep down predators in the thinly populated district around Alveston, where there had been plenty of woods in which the birds could nest, the situation had changed. Knapp said that the clearing of woodland and intensive game-keeping meant that, 'our losses by such means [vermin] have become a very petty grievance.' This was the perception of disinterested observers over much of the country. Anderson quotes a contemporary who said that payments ceased in his parish after the New Poor Law of 1834. This law was designed to force the poor to accept indoor relief in the workhouses; the vermin payments were presumably halted to remove suspicion that they were a covert form of outdoor relief which might enable n'er-do-wells to go on scraping a living of sorts and wander about the pheasant coverts in the process.[xxvi]

Some farmers had always paid privately for the killing of pests. William Cobbett, who was not inclined to be free with his money, agreed a generous tariff with a man at Botley, Hampshire, and certainly paid him in 1809.[xxvii] A wide scatter of evidence shows that the practice continued after the parishes had ceased to pay but just how widely is uncertain. Well into the twentieth century, farmers encouraged local men and boys to trap pest species. It was a form of sport in villages where entertainment was lacking in the winter evenings. The youths bat-fowled along the hedges after dark and swept the roosting birds, even sparrows, into their nets. The catch went into 'vermin pie', which was acceptable to poor families when protein in their diet was scarce — badger digging, now a shadowy criminal sport, was carried on into the second half of the twentieth century as much for the sake of the hams as for devilment. From the farmers' point of view all this trapping, digging and killing was pest control for free.

Antiquarians have compiled the parish vermin data for some counties. Population trends can be discerned in a few cases, though would be desirable to construct an independent index of the rewards offered, since otherwise one cannot tell whether what are being observed are changes in pest numbers or in incentives to participate. House sparrows, pests on cereals, went on being killed in large numbers, although we may surmise that the mighty effort at reducing them was little more than 'rat-farming'. That is to say, the survivors of each onslaught were left with the best nesting sites and supplies of food, and soon replenished their population. At the other extreme, a few species vanished totally from lowland England, notably red kites and buzzards. We have to assume they were subject to the opposite effect, the Allee effect, whereby survivors became too few to find mates and their numbers faded away.

The decline of raptorial birds may have been as much the consequence of the persecution of predators by game-keepers, who grew very numerous during Victoria's reign, as it was of premiums paid by the parishes for vermin. Both forces operated, though the most intensive phases did not overlap for long, and game preserving by itself was capable of eliminating, say, buzzards from lowland England. Much-reprinted maps show the inverse relationship between game-keeper and buzzard numbers, while the rebound of the buzzard population now that estates buy in young pheasants and attend less to 'pest' control is a backhanded tribute to the former consequences of preserving game.[xxviii] The reality is that buzzards take few pheasant poults. But the vocal game-bird industry presses for public money to be spent on 'controlling' the species in its own private interest.

Finally, the distinction between species that survived despite intensive efforts at control may have been between those whose breeding followed a K-strategy, i.e. producing a small number of offspring in which parental investment was high and where mortality was expected to be low, as opposed to an r-strategy, i.e. producing large numbers of offspring and accepting high losses at young ages. Large raptors fell into the former category, which made them vulnerable to what might be called over-harvesting by game preservers; house sparrows fell into the latter category and survived every effort at reducing their numbers.

ENDNOTES

[i] M. Ambrosoli, *The Wild and the Sown: Botany and Agriculture in Western Europe: 1350–1850* (Cambridge: Cambridge University Press, 1997).

[ii] E. L. Jones and S. J. Woolf (eds.), *Agrarian Change and Economic Development: The Historical Problems* (London: Methuen, 1969), p. 4.

[iii] C. Lane, 'The development of pastures and meadows during the sixteenth and seventeenth centuries,' *Agricultural History Review*, **28** (1980), pp. 18–30.

[iv] T. Williamson, *The Archaeology of Rabbit Warrens* (Princes Risborough: Shire Publications, 2006).

[v] M. Warriner, *A Prospect of Weston in Warwickshire* (Kineton: The Roundwood Press, 1978), p. 80.

[vi] R. E. Baldwin, 'Patterns of development in newly settled regions,' *Manchester School of Economic and Social Studies*, **XXIV** (2) (1956), pp. 161–179.

[vii] J. G. Cornish, *Reminiscences of Country Life* (London: Country Life, 1939), pp. 1–3.
[viii] C. C. Taylor, 'Whiteparish,' *Wiltshire Archaeological and Natural History Magazine*, **62** (1967), pp. 79–102.
[ix] P. Brandon, *The Sussex Landscape* (London: Hodder and Stoughton, 1974).
[x] R. A. D. Cameron and P. J. Dillon, 'Habitat stability, population histories and patterns of variation in *Cepaea*,' *Malacologia*, **25** (2) (1984), pp. 271–290.
[xi] W. G. Hoskins, *Provincial England* (London: Macmillan, 1963).
[xii] E. Power, *Medieval People* (Harmondsworth, Middlesex: Penguin, 1937), pp. 141ff.
[xiii] J. Steane, *Oxfordshire* (London: Pimlico, 1996), p. 40.
[xiv] L. G. R. Naylor, *The Malthouse of Joseph Tomkins, 58, 60 East St Helen's Street: The Story of an Abingdon House* (Privately published).
[xv] N. Hammond, *Rural Life in the Vale of the White Horse 1780–1914* (Newbury: Countryside Books, 1974), p. 55.
[xvi] E. L. Jones, 'The Land that Richard Jefferies Inherited,' *Rural History*, **16** (2005), p. 86.
[xvii] Williamson, *Rabbit Warrens.*
[xviii] J. Sheail, *Rabbits and their History* (Newton Abbot: Country Book Club, 1972).
[xix] R. J. O'Connor and M. Shrubb, *Farming and Birds* (Cambridge: Cambridge University Press, 1986), pp. 5–7, 91–92.
[xx] N. W. Moore, *The Bird of Time: The Science and Politics of Nature Conservation* (Cambridge: Cambridge University Press, 1987), p. 29.
[xxi] C. Aspinall-Oglander, *Nunwell Symphony* (London: The Hogarth Press, 1945), p. 23.
[xxii] G. Blainey, *A Land Half Won* (South Melbourne: Macmillan, 1980), p. 313.
[xxiii] E. C. Rolls, *They All Ran Wild: The Story of Pests On the Land in Australia* (Sydney: Angus and Robertson, 1969), p. 78.
[xxiv] E. L. Jones, 'Bird Pests of British Agriculture in Recent Centuries,' *Agricultural History Review*, **XX** (2) (1972), pp. 107–125; P. J. Dillon and E. L. Jones, 'Trevor Falla's Vermin Transcripts for Devon,' *The Devon Historian*, **33** (1986), pp. 15–19.
[xxv] J. Knapp, *The Journal of a Naturalist* (London: John Murray, 1829), p. 218.
[xxvi] D. Anderson, ' "Noyfull fowles and vermin": Parish payments for killing wildlife in Wiltshire, 1533–1863,' *Wiltshire Archaeological and Natural History Magazine*, **98** (2005), p. 151.
[xxvii] W. Cobbett, *Farm Account Book 1810–12*: Nuffield College, Oxford, MS. XIII.
[xxviii] For the maps, see C. R. Tubbs, *The Buzzard* (Newton Abbot: David & Charles, 1974).

5. AGRICULTURAL CHANGE: SOUTHERN ENGLAND

The stone-curlew seems generally to have increased in this down-land country during the last fifty years, but now owing to the re-cultivation of the waste and abandoned land, it is tending to decrease

George Brown (1948)

AGRICULTURAL CHANGE

Agriculture is a primary means by which human needs are imposed on the environment and find expression in the landscape. In England, the Parliamentary enclosure movement of the late eighteenth and early nineteenth centuries is typically regarded as a signal influence on the density of birdlife, and by extension that of other creatures, though with little direct evidence. Almost no other changes in farming are given such a central role: most are given no role at all and it is therefore appropriate to start with the enclosures as constituting a major break in ecological history. In the midlands, which the great belt of enclosures covered from southwest to northeast, the planting of hedgerows around the large fields imposed on the formerly open terrain of strips undoubtedly brought about an increase in song birds. Hedges supplied shelter and nesting sites. The corollary is that small birds would have been far scarcer in the windy, open fields of medieval parishes — an assumption which eliminates romantic notions of a timeless England.

The argument is made by analogy, pointing to the lack of hedges and accompanying paucity of hedgerow birds in the type specimen of an open-field parish that survives at Laxton, Nottinghamshire. Similar contrasts between open and enclosed landscapes are visible in a few other places, and an American example offers a more distant analogy. In the workaday Amish country of Pennsylvania, where the Pennsylvania Dutch still farm with horses, there are no hedgerows, and few passerines are to be seen.

Hedges had a positive value for farmers besides demarcating the bounds of private property. They acted as stock fences and shelters for livestock. Elm trees could be grown at intervals along them to provide wood for farm purposes, such as resilient plank floors for carts and waggons (elm dents rather than splits). From a high vantage point, nineteenth- and twentieth-century landscapes came to look as if they were completely wooded, until the elms were subtracted by the Dutch elm disease of the 1970s. But hedges had the disadvantages of overshadowing crops and harbouring pests. They required annual cutting if they were not to encroach on the cultivated area. Modern work suggests that the popular mantra of dating hedgerows by the number of species of shrubs (the more, the older) is too simple, but it is clear that some hedges are genuinely ancient, as distant in their origins as Saxon times. Saxon or not, enclosures were established early in some parishes, well before the late eighteenth century created the hedged landscape thought typical of the English countryside. Nor was enclosure the sole origin of the hedged landscape. The western and eastern flanks of England were carved straight out of the forest by the process known as assarting. These areas had rarely been in open field and on the western side of the country always retained plenty of wooded cover.

At least as great in their impact as enclosures were drainage and reclamation, if the under-draining of fields is included. Drainage has chiefly attracted attention because of the conspicuous water birds it drove out. Species once present in the Fens of East Anglia included cranes, avocets and ruffs, which ceased to nest as the waters were dried. Marshland was increasingly circumscribed by drainage, but in the nineteenth century, gunners and taxidermists could still make a living by selling rare specimens to the collectors of stuffed birds, while decoys could be used to trap duck for the market. These trades were on a commercial scale, they are reasonably well documented, and have inevitably scandalised conservationists and historians of ornithology. We will deal with the impact in a later chapter.

Drainage extended the arable area but the advance of the plough into hitherto uncultivated land was a more general force. Apart from certain notorious periods of falling grain prices, and hence of arable retreat, reclamation was remorseless. Each advance overran the limits of previous expansions. It is difficult to map them comprehensively because, while there were some highly visible ploughing campaigns, reclamation also took place in almost covert ways. A scatter of references does exist to the practices by small holders of 'rolling hedges' and cultivating surreptitious bites from the edges of moorland, heath and forest.[i] The references tend to be obscure and most historians have shied away from the labour of reconstructing the creeping conversion into farmland of the 'waste', which was not really unused but employed for low-intensity grazing on poor quality herbage and for gathering any other natural product that could be scrounged.

A SHIFTING BASELINE

Ample information is available concerning the Parliamentary phase of enclosure. As usual, where data can be readily tabulated, scholars buzz around the topic like bees at the hive. Intense interest in the political and institutional aspects of re-allocating land has made it central in English social history. Yet the visibility of the process may be a little out of proportion to the ecological impact. What is at stake is not enclosure as such, which took various forms involving property rights, not all of them with the same consequences for the landscape. The environmental essence was converting pasture or the waste to ploughland or improved grassland, processes following at varying intervals after enclosure, or even occurring independently of it. Shrubb observes that many species formerly present on the waste adapted reasonably well to the farmland which replaced it, *but in lower numbers.*[ii] This is a pregnant conclusion, an instance that is consistent with Frans Vera's hypothesis of a shifting baseline in which each ecosystem was poorer than its predecessor.

A much-discussed example of a species lost to reclamation was the turkey-sized great bustard. The bustard was so conspicuous that every last specimen (*le mot juste*) was listed in the appropriate county avifauna. The birds were driven out of the Wessex chalk-lands by disturbance and shooting in the early nineteenth century. Other species declined as the open downs were reclaimed, but bustards and birds of prey, like hen harriers, were conspicuous and 'privileged' by observers above lesser losses and certainly above species whose range or population actually expanded when new acres were brought into cultivation. The phenomenon is familiar from experimental economics: people lament what they lose more than they celebrate a matching gain. Among the losses, the bustard was what would be called in the modern cant an 'iconic' species and its loss was recognised at the moment — precisely at the moment — when it vanished from southern England. A. C. Smith, author of *The Birds of Wiltshire*, was already hankering in the nineteenth century for the open downland of the century before his own; it almost makes one sympathise with Vera's thesis.

As soon as the *Wiltshire Archaeological & Natural History Magazine* was founded in 1854, Smith began quizzing its readers, 'can you glean any particulars of the now extinct Bustard, from old inhabitants of the Plain, shepherds, labourers, farmers, and others, who have been eye-witnesses of the bird in a wild state?'[iii] A response came smartly from J. Swayne, who said that he recalled farmers near Lavington talking about the scarcity of bustards on the down in 1785 or 1786, 'which they attributed to the heath, &c., being broken up and converted to tillage, and to the corn being weeded in the spring, whereby the birds were disturbed and prevented from making their nests'.[iv] Swayne himself had seen bustards in the 1780s and reported a few that were captured afterwards, when they were becoming very scarce. Gilbert White collected similar reports in Hampshire. Observations trail away in the early nineteenth century, then cease, and later history is a sad tale of stragglers until the reintroduction of birds from Russia at the start of the twenty-first century.

The environment of light land areas was transformed. The planting of copses and shelter belts, to harbour game or foxes more than benefit agriculture, had a striking visual impact. Trees broke up the expanses of grassland on which the bustard lived. On the other hand, the new spinneys and belts were inadvertently suitable habitat for quite different species. The effect was outstanding with respect to the wood pigeon, which thrived because it liked the ecotone (where different habitats meet, in this case along the edges of linear woods) for nesting, roosting and keeping watch. Pigeons found crops in the new rotations tempting and their numbers grew to be a plague. Mid-nineteenth century farmers on the Cotswolds ceased to grow vetches because pigeons ate them bare. Details of repercussions on other species are largely unreported — not everything increased to the same extent as wood pigeons or attracted attention as pests. But here and there it can be seen that other species did adapt gleefully to the new environment. Whereas the bustard shrank into oblivion, less prized birds became commonplace.

As the mixed farming — high farming — of the mid-nineteenth century intensified, it thus created its own problems. The system was an interlocked one of high inputs and high output; agriculture was run on the lines of industry, and farms could be described as roofless factories. Mixed farming occupied a distinctive landscape, even if the products were varied. Stereotypical high farming centred on close-knit rotations of which the best known is the Norfolk four-course, two 'white straw' crops (wheat and barley) alternating with two fodder crops (turnips and clover). The fodder crops were 'fed off' by sheep and cattle, whose dung was returned to the soil to fertilise the following cereals. This was a self-contained arrangement in which the by-products of one enterprise became the inputs of the next, while the array of products helped to guard against prices falling for any one of them.

The repetition of the same crops imposed an ecological penalty. Dense crops encouraged pests and pathogens. Sown and sown again, the turnip crop was liable to suffer from finger-and-toe disease and the clover to suffer from 'clover sickness.' The problem may be judged from the fact that the only remedy for clover sickness was a sort of scorched earth policy, refraining from sowing the crop on affected land for as long as five or even eight years. Rotations had to be elaborated so that mixed farming's intended results could be secured by round-about means. Was high farming therefore approaching some Kjaergaardian climax when ecological calamity would force its abandonment? Although the system did show signs of strain from pathogens, predators and weeds, with its costs swelling, it was probably not truly close to disaster. But before the experiment could come to any such fatal conclusion, it was aborted when cheap grain came flooding in from the New World during the 1880s, whereupon much of the downland was farmed less intensively or tumbled down to weeds and scrub or was turned back to grass.

Ploughed land that favoured seed-eaters contracted during the depressions of the late nineteenth century and the inter-war years of the early twentieth century. Tumbledown pasture took over. A report to the Royal Commission on Agriculture in 1894 found that, 'in every direction from Andover, uncultivated land may speedily be met with, either corn stubbles never broken up and allowed to grass themselves, or old "seeds" or sainfoin layers remaining down long beyond the usual time'.[v] Although this did not revive the bustard, it made an opening for rabbit populations to swell and nibble the sward close and tight, permitting orchids to grow and favouring assemblages of grassland birds. They found it hard to cope with the subsequent growth of coarse grass, let alone the resumption of ploughing whenever cereal prices returned to being high.

The largest species favoured by tumbledown land was the stone curlew, another large bird of open country, though not so bulky as the bustard. The stone curlew continued to breed throughout the ploughing campaign of the Second World War and into the post-war period, becoming the bustard's symbolic equivalent. Metaphorically speaking, Vera might regard the stone curlew as an inferior substitute for the bustard, though it had been present all along.

Later, naturalists were nostalgic for the 11 stone curlews' nests that the late George Brown found in 1937 in one small area close to where Hampshire, Berkshire and Wiltshire meet. They envied anyone who had seen the autumn flocks of 200 or more that the beaters flushed when they tramped through the kale to put up partridges for the guns.[vi] Brown described the terrain he studied as consisting of farmland abandoned during the 1930s depression and grown over with rough grass and stunted blackberry bushes. The area had become rabbit warren, with patches of flints strewn among poor grassland, dotted with occasional gorse bushes and seeming like a desert to eager farmers. But after 1940, almost all the land was once more reclaimed for cereal-growing. Breeding stone curlews, which had increased during the depression, fell steadily. By 1948, there was only half the number of 10 years earlier.

The decrease was not an immediate response to ploughing. The birds gave every sign of taking readily to fields of young oats and barley. But chicks and young birds did not like wheat fields, which Brown thought was because wheat holds dew and rainwater. A greater problem was that the flints under which adult stone curlews found their food were being constantly turned over by implements, especially in the spring. Not enough birds may have been able to find food to sustain a breeding population. The rolling of the fields in April and early May often destroyed the first brood of any pair that did manage to nest. Yet in the 1950s, stone curlews were not really rare on the Wessex chalk as a whole. In summer, I could show interested visitors several pairs, and in autumn, the beaters could flush a score or two at a time. The species had seemingly adjusted to nesting on cultivated land, though at a lower density than on downland grass. The ultimate outcome was different: ever more intensive farming methods distressed the bird, its status was on a knife-edge, and in the end it failed to adapt.

TRAPPING

What happened, or did not happen, to wildlife on the extended arable is interesting. An account by Shrubb suggests that the impact of habitat change during the nineteenth century was less than its scale might suggest.[vii] His main point is that the progress of enclosure was spread over more than a century (in reality it was longer), allowing time for animals and birds to adapt. There may have been lags before some species changed their behaviour and began to recover their numbers: Victorian bird books show that in the breeding season linnets moved back from the new farmland, where they wintered on spilled grain, in order to nest in their old habitat of gorse on the wastes. They did not usually adapt to nesting in the hedges of the new farmland until the late nineteenth century.

By contrast, the goldfinch never did adapt boldly to the intensive 'high farming' environment, despite the benefits other seed-eaters found in the varied food then present (and so lacking on the arable prairies of late twentieth- and early twenty-first century England). To the goldfinch, the stack yards that nineteenth-century sources insist were important to most seed eaters were not attractive. Goldfinches depended on a narrower range of weed seeds than other carduline finches and, as Shrubb points out, the thistles on which they fed were declining. Another weed important in their diet, groundsel (*Senecio*), was becoming rare under the intensive weeding regime made possible by the cheap labour of the high farming era.[viii] Bird catchers also took astonishing numbers of goldfinches to sell to town customers as cage-birds, but by late Victorian times the days of plenty, when a boy could trap forty dozen in a morning on the Sussex Downs, were over.[ix]

Finches were favourite prey for the bird catchers, although these men would capture anything for which they had a commission or could hope to sell.[x] They often had orders for birds to be used as targets in trap-shooting, banned now though scarcely crueller than the shooting of reared pheasants. A man from Bristol used to arrive at Theale, Somerset, with his trap full of young wild birds to release for the local farmers to shoot competitively while drinking deeply of cider.[xi] Nets were thrown over the hedges to capture roosting birds, which were released next morning for more shooting. During the nineteenth and early twentieth centuries, everything was fair game. The minutes of a club of Cambridge students record in 1855 that the members, 'fell in with a flock of rare linnets and shot about 50. Lunched, adjourned to the fen and killed four owls'.[xii]

Goldfinches and linnets were taken for the cage-bird trade because of their pretty looks and sweet songs. They were taken in clap-nets to which they were decoyed by trained call-birds ('brace birds') or stuffed specimens ('stales'). In about 1860, 130,000 goldfinches were caught every year near Worthing. Goldfinches were partial to thistles and Norfolk bird catchers kept thistle heads in brown paper in dark cupboards, ready to bring out for winter trapping. Despite this evidence of trapping, the fortunes of goldfinches and linnets probably fluctuated more in the long run through changes in their habitat. Both haunt waste ground and gorse commons, uses that are squeezed today between expanding plough-land and the growth of woodland for which the public's unthinking 'green' admiration is, as Mark Cocker says in *Birds Britannica*, 'obsessive'. In the 1870s, a decrease in the number of goldfinches was being put down to the expansion of arable and its greater cleanliness.

Caged song birds were almost universally admired before the second half of the twentieth century. Henry Mayhew, author of *Life and Labour of the London Poor*, visited a rat and bird catcher who kept linnets in square boxes scarcely bigger than bricks. 'Toy' linnets — they were treated as toys — learned to widen the range of their song by copying other species. They fetched high prices, whereas 'broken' song birds that did not sing well were cheap.[xiii] Bird catchers did the rounds of rural areas throughout the country. An old trapper from Cheltenham toured the Vale of White Horse annually in the 1880s and 1890s, buying birds caught by the villagers, which suggests that trapping by professionals understates the pressure on birdlife.

At one time, bird catchers in Gloucestershire might take from 5 to 14 dozen goldfinches per day and flocks of 200 to 300 were to be seen, but no longer. By 1902, the numbers were diminishing in the county, although they may have risen a little in the two or three years before that, thanks to county council protection. National demand really shrank only in the face of competition from imported canaries and budgerigars, and most of all, much later, when alternative entertainment in the home became available from radio and television. This led purchases to fall off far faster than humane and conservational impulses might have managed on their own. (Making collections of birds' eggs also ceased when television appeared. Although the Wild Birds' Protection Act of 1954 helped in England, television was probably more important — its advent coincided with an abrupt cessation of egg-collecting in Australia too).

Professionals took birds in the summer by means of the 'water-trap', which amounted to no more than netting a long list of species when they came down to drink from brooks or ponds in hot weather. Otherwise nets were set for finches, and brace-birds were stationed on 'flurr' sticks that were agitated to make them flutter. This drew in wild birds; no bait was needed. Methods and apparatus were documented by the early bird photographer, Richard Kearton, when large-scale trapping was drawing to a close at the end of the Victorian era.[xiv] Five dozen linnets might be taken at a time during the 'November flight of linnets' between 15 and 30 November.

These were autumn migrants arriving at the crack of dawn, supposedly from Scotland, though modern observations suggests they also came from northern Europe via Holland. Prices fluctuated markedly, though no long-term diminution was recognised. In 1878, the birds were scarce enough to send the trappers scurrying to Ireland, sending a goodly number of linnets back to London. This glutted the market and the price dropped by as much as 60%.

Another regular means of capture was bird-liming, trapping the birds on a stick smeared with a sticky substance. Bullfinches were taken this way. A 'call bullfinch' was set down in a cage to lure wild ones into settling on the limed twigs. Two dozen might be taken in a day, though one suspects this was in market gardens or fruit-growing districts where bullfinches concentrated and the trapper probably moved around the area. Goldfinches, chaffinches and woodlarks were also captured by bird lime.

Entire breeding groups of linnets were wiped out near London, Glasgow and Edinburgh and populations in eastern England were reduced by the catching of migrants near London. British migrants were taken in large numbers in Belgium, France and Spain. Until the early 1870s, a score of linnet catchers worked the dunes near Great Yarmouth. When they faded away, it was less through decreasing demand than the opportunity of more regular employment or reduced numbers of birds to catch, or both.[xv] Similarly, goldfinches were rendered extinct or nearly so in northern England and southern Scotland by the 1870s. Their numbers were low on the Norfolk coast by the start of the twentieth century. Successive protective Acts passed during the 1890s had scarcely been enforced and the catching and confinement of British songbirds was not made illegal until 1953. Nevertheless the new mood cramped the style of the bird catchers, and linnets, at least, underwent only local fluctuations in numbers during the first half of the twentieth century. It was during the second half of the century that more conventional agro-environmental forces — the loss of weed seed through intensive herbicide use — further, or once again, depressed populations.

The finches illustrate why we should not assume that changes in land use automatically 'post-dict' what happened to wildlife, however patent the changes may seem and whatever our modern experience of habitat preferences leads us to expect. All species do not respond in the same way or to an equal extent. The issue needs to be tackled on a species by species and period by period basis. Unfortunately, data to permit this are at best limited, though occasional examples may be found. The house sparrow, much more aggressive than linnet or goldfinch, was far quicker at adapting to new environments. John Knapp noted that the sparrow became 'immediately an inhabitant of the new farm house, in a lonely place or recent enclosure', and took almost every ear of grain yards in from the edges of fields.[xvi] Around towns, where sparrows were plentiful, long-awned wheat had to be sown to make it hard for birds to perch and eat the ears.

Available observations vary greatly among species. Some attracted plenty of attention from naturalists, writers and poets but small, quiet, retiring, cryptically coloured species hardly figure in Victorian literature or science. Of field birds, the skylark is conspicuous in the record (highly audible songster, favourite human food, sometimes occurring in pest proportions), but the meadow pipit qualifies on none of these grounds. Even so, although annual sales of skylarks in the markets of Victorian London ran to the hundreds of thousands and farmers persisted in shooting them as harmful to growing cereals, field observations in the past were often little more than rhapsodies to the bird on its song flight.

The experience of the skylark is worth closer examination, though the numbers reported as taken are hard to interpret — or believe.[xvii] The bird was netted around two chief centres, Dunstable and Brighton. Some 50,000 were taken on the Dunstable Downs in a single winter. In Victorian times, about 215,000 from the two counties of Cambridgeshire and Bedfordshire were sold at the Leadenhall market every year, two thirds of London's consumption. In the latter part of the century, Great Eastern trains bound for the capital were sometimes filled with skylarks, sacks of 20,000 or 40,000 day after day.

It is best to look at the matter from the London end, since the markets there received birds from all the Home Counties, as well as part of the catch from the South Downs near Brighton. Mayhew thought the annual catch in 1851 was 60,000 'larks' — probably not only skylarks. In 1854, the number of skylarks entering the London meat markets alone had risen to 400,000 per annum. If prices dropped, some were sold as cage birds, this probably being a steadier trade; they were often blinded because this was supposed to make them sing better. The Brighton area also supplied London but exported as well to the great Paris market of Les Halles. In the winter of 1867–1868, 1,255,000 were sent to Dieppe. London poulterers were glutted with them in the 1890s, when they were destined to be savouries at club dinners. Trapping skylarks for sale continued in England until they were protected by law in 1931, although as recently as 1997 falconers were permitted to take some hundreds every year with merlins. This was an unusual example of a practice long permitted to continue, contrary to the spirit of bird protection, on grounds of cultural tradition. On the continent, protection has not yet been extended to skylarks and they are hunted under the EU Birds Directive, the kill remaining massive in the twenty-first century.

Another species trapped in great numbers for food, especially as a migrant on the South Downs, was the wheatear — a little butterball it was called. The fact raises a thorny problem about the net effect of killing birds. Was trapping like this only, or mainly, taking a surplus that would have been lost anyhow, through natural causes, before the next breeding season? Annual catches at the levels recorded could not have been sustained had there not been an element of harvesting redundant population. On the other hand, we noted that goldfinch populations seemed unable to bear the Victorian traffic in them.

Did birds and other wildlife really decline overall? A problem arises from biased observation. Wildlife records derive disproportionately from shot specimens, notes by gunners, lists of gamekeepers' gibbets and museums full of avian corpses, stuffed and laid out in trays or mounted and displayed in glass cabinets. Even when they were not themselves collectors, nineteenth-century ornithologists were drawn to the more visible species, while those that became scarce attracted their special notice, at least if they were game birds or raptors. There was competition to collect the rarest specimens. The rarer the better, a sort of cloud forest syndrome, which means we have more provenanced records of strays than of everyday birds.

Observations of common species were made *en passant*, when they were made at all; were seldom based on actual censuses; and were biased because they were ones that lived close to human settlements or possessed bright colouration, complex songs, palatability or roles as pests. This gave a wide range of possibility and indeed there was nothing that might not be shot or trapped in Victorian times, but even then some species were far more attractive as targets or pets than others. Tawny owls figured conspicuously. Men living in Savernake Forest, Wiltshire, were highly skilled at lassoing the tawny owl as it left hollows in trees, and from Bablock Hythe, near Oxford, a specialist owl-finder went all over the countryside catching owls or taking them from the nest. He carried a wicker-work 'chair' with three or four compartments for his catch, which he tamed and sold to gardeners (who preferred them to cats for controlling vermin) or to the gentry as pets.

We need not continue; we could never make a complete record of all the ways in which birds were exploited. For the nineteenth century, and sometimes earlier, anecdotes like those about tawny owls are easy to find. It would not be fair to call them entirely fungible because each species is different and circumstances varied. There are ample records of the exploitation of some (skylarks) and few of others (meadow pipits). Yet there really was a degree of substitutability, because many species were affected by similar changes in land use and trends in human predation.

When human populations became more citified, links with the wild were weakened. This did not mean the pressure on wildlife was much reduced, often the opposite. City populations were richer, slums notwithstanding, and formed vast markets for food and raw materials. The pull of the market extended from London to rural areas, drawing in every type of biological commodity — anything that might be eaten and an astonishing array of items that could be used to ornament or adorn. Mayhew found sellers of birds' nests on the London streets. The nests were displayed as ornaments, and it is a well-known story that the great crested grebe was reduced to 42 pairs in the country by 1860 as a result of the female fashion for using its head feathers as tippets.[xviii]

The tippets had been imported from grebes slaughtered overseas, chiefly in southern Europe, as an instance of the international trade in wildlife and body parts. Then, in 1851, a foolish letter in *The Zoologist* reminded people that the species could be taken in England and that its plumage was a beautiful substitute for fur. The assault turned inwards and the population plunged. Thomas Southwell thought the great-crest might become extinct; its eggs were always taken when found and he had known 30 or 40 gathered from a single Norfolk Broad.[xix] Bird protection laws followed in the 1870s, with the great-crest in mind. They were flouted, as usual. As late as 1931, the species was still being shot for the taxidermists and to reduce damage to fishing interests (little grebes and herons were still shot for that reason much later). Yet the laws of the 1870s, coupled with protection on some private waters, did have an effect: Even Southwell admitted that the species had not decreased in the years leading up to 1879, which he attributed to the Sea Birds Protection Act and the tax on guns, which had presumably disarmed the worst of the hobbledy-hoy shooters who shot this type of prey because they could not afford to shoot game birds. It is customary to restrict the poor man's sport and wink the eye at the rich man's.

By 1907, the fashion of wearing tippets had died its own death, and the main reason for exploiting the great-crest evaporated. A census of 1931, an early example of mass participation by bird-watchers, counted a population of very close to 1,240 pairs. This shows how elastic a recovery may occur in a fortunate species (unhappily, not in all) once exploitation is curtailed. By the early twenty-first century, the population was about 6,000 pairs but this further growth was substantially due to land-use changes that happened to favour the species: the post-war digging of gravel was filling the river valleys with flooded pits suitable for feeding and nesting.

CONCLUSIONS

On the face of things, the declines of the bustard and stone curlew fit the shifting baseline hypothesis which sees the environment as always running downhill. Yet the deterioration is our choice and does not always happen. Studies of the large block of grassland on Salisbury Plain, a military training area not greatly disturbed by farming, show it to be a vast refuge for certain species — like the stone curlew — absent or rare on surrounding agricultural land. Salisbury Plain was colonised by new farms in the mid-nineteenth century, not long before the depression reduced prices and the War Office acquired the Plain, letting the land revert to grass. Most of southern England is cultivated today and is far less exciting to the naturalist than the inadvertently 'historical' ecosystem of the Plain. It is on Salisbury Plain that conservationists have reintroduced the bustard.

The pauperised birdlife of southern English farmland is a result of modern intensive farming, which is a purely political construct — subsidies under the Common Agricultural Policy protect farm incomes at prodigious expense to the taxpayer. If the protection of farmers in the European Union were overcome, this situation might be reversed. Not every grassland species would recover to the same extent; we have already noted the differences, ultimately genetic, that prevent a perfect match with land-use changes. But many would do so, as the rich downland birdlife of Salisbury Plain demonstrates. We need not have only a descending staircase and conservationists need not be reduced to fighting rearguard actions.

Assuming the goal is to maximise people's satisfaction in observing wildlife, there is no great environmental problem, just a political one. The removal of taxpayer support for agriculture would let English farming quit marginal areas, which would be colonised again by wildlife. 'Farmers' reaction to this seems to be the curious argument,' observes Shrubb, himself a farmer, 'that large areas of the countryside would then revert to wilderness, offered as a sort of bogeyman to frighten us'.[xx] It is astonishing to find the farmers' view paraded by Sir Colin Spedding, who urges that agriculture preserves the countryside in an attractive and usable form and if the land is not farmed it will go to scrub and woodland.[xxi] Attractive to whom and usable by whom are good questions. I cannot do better than repeat Shrubb's riposte: 'what is wrong with that, it's the whole point'.

Ironic though it may be, land use is a product of prices and taxes and can be altered accordingly. Permanent loss, complete extinction, is rare. In large measure, society can choose to 'farm' whatever mix of wildlife it prefers but must accept the trade-offs. For instance, an implication of the shifting baseline thesis is that the longer the species list the better. This is a maximum biodiversity fallacy. Subjective quality counts and less may be more; if not, temperate zone naturalists would move to the lush tropics. The 'stamp collectors' among them might actually wish to do so but others prefer the astringent, aesthetic quality and historical associations of their temperate zone homelands. Choice is inevitable but crisis is not and the baseline need not shift only downwards.

ENDNOTES

[i] W. G. Hoskins, 'The reclamation of the waste in Devon, 1550–1800,' *Economic History Review*, **XIII** (1943), pp. 80–92; N. Hilton, 'The land's end peninsula: The influence of history on agriculture,' *Geographical Journal*, **CXIX** (1953), pp. 57–72.

[ii] M. Shrubb, 'Farming and birds: an historic perspective,' *British Birds*, **96** (2003), pp. 158–177.

[iii] A. C. Smith, 'On the ornithology of Wilts,' *Wiltshire Archaeological and Natural History Magazine*, **I** (1854), p. 43.

[iv] J. Swayne, 'The bustard,' *Wiltshire Archaeological and Natural History Magazine*, **II** (1855), p. 212.

[v] Quoted in J. Sheail, *Rabbits and their History* (Newton Abbot: Country Book Club, 1972), p. 140.

[vi] G. Brown, 'The stone curlew,' *Transactions of the Newbury District Field Club*, **IX**, No. 2–4 (1951), pp. 26–32.

[vii] M. Shrubb, 'Farming and Birds,' passim.

[viii] M. Shrubb, 'Farming and Birds,' p. 167.

[ix] E. S. Turner, *All Heaven in a Rage* (London: Michael Joseph, 1964), p. 194.

[x] F. Buckland, *Notes and Jottings from Animal Life* (London: Smith, Elder, 1894 edition [original 1882]); M. Cocker and R. Mabey, *Birds Britannica* (London: Chatto & Windus, 2005); J. G. Cornish, Bird notebooks, three

volumes, 1883–1912, examined at the Wantage Field Club; S. Holloway, *The Historical Atlas of Breeding Birds in Britain and Ireland 1875-1900* (London: T and A D Poyser, 1996); H. Mayhew, *London Labour and the London Poor* (London: Griffith, Bohn & Co., 1861); and W. L. Mellersh, *A Treatise on the Birds of Gloucestershire* (Gloucester: John Bellows, 1902).

[xi] Theale Millennium Committee, *Theale Past and Present* (Privately printed, 2002), p. 104.

[xii] Quoted in *Shooting Times*, 8 June 2011.

[xiii] For descriptions of cruelty, see Turner, *Heaven*, pp. 198–200.

[xiv] R. Kearton, *With Nature and a Camera* (London: Cassell, 1898) [Glasgow Digital Library].

[xv] A. H. Patterson, *Nature in Eastern Norfolk* (London: Metheun, 1905), p. 76.

[xvi] J. Knapp, *The Journal of a Naturalist* (London, 1829), p. 218.

[xvii] P. F. Donald, *The Skylark* (London: T. & A. D. Poyser, 2004); E. S. Turner, *Heaven*, pp. 195–196; M. Cocker and R. Mabey, *Birds Britannica.*

[xviii] M. Cocker and R. Mabey, *Birds Britannica*, pp. 6–7.

[xix] R. Lubbock (new edition by T. Southwell), *Observations on the Fauna of Norfolk* (Norwich: Jarrold and Sons, 1879), p. 122, n. 122.

[xx] M. Shrubb, *Birds, Scythes and Combines: A History of Birds and Agricultural Change* (Cambridge: Cambridge University Press, 2003), p. 328.

[xxi] C. Spedding, 'Why the Farming Industry is Vital to Us All,' *Hunting*, **14** (126) (2007), pp. 48, 50.

6. LANDSCAPES OF DESTRUCTION: THE CURSE OF THE PHEASANT

> I'd rather that my son should hang than study letters. For it becomes the sons of gentlemen to blow the horn nicely, to hunt skilfully and elegantly, to carry and train a hawk. But the study of letters should be left to the sons of rustics.
>
> Quoted by Mark Girouard, *Life in the English Country House* (1980)

Killing birds and animals is a preoccupation for large numbers of those who can afford it. Landowners have the advantage. Limited access to land confers prestige on the activity, making the slaughter of wildlife a badge of status or social acceptance. Consider the early eighteenth-century movements of the Imperial Austrian court; they were determined solely by the hunting seasons.[i] Hunting ceased only for Lent, until woodcock shooting began at the end of March. April was spent flying falcons at herons, just when the herons were making beelines for their nests. From the end of June until the start of October, the court moved elsewhere to hunt stags, decamping again for 10 days shooting of hares and pheasants. After that, it was engaged almost every day in hunting wild boar, until the end of January, when enormous drives were staged to deal with any game that had escaped. At a later period, the Archduke Franz Ferdinand managed to slaughter 274,889 animals and birds before he fell to the assassin's bullet at Sarajevo.[ii] He recorded killing his 100th eagle, 1,000th chamois and 6,000th stags before the end of his abbreviated lifetime.

Rulers and grandees have always been eager participants in shooting (hunting in American English). Political affiliation is irrelevant. One after another, the Kaiser, Hermann Goering (described by Wikipedia as 'a lovely man'!) and Erich Honecker all shot on the same estate outside Berlin. Tribunes of the people, such as Tito and Ceausescu, neither of them tender towards human adversaries, were also embroiled in the competitive slaughter of birds and mammals.

In Western democracies, shooting continues to engage the elite, for example, Prince Phillip in the United Kingdom, Prince Bernhard of the Netherlands (despite the fact that he was President of the World Wildlife Fund), and more recently, King Juan Carlos of Spain. President Sarkozy entertained Col. Gaddafi at a shoot and proposed to re-start the Chasse Presidentielle, until frustrated by a French economy drive.[iii] (The Chasse Presidentielle involved the palace guard, a torchlight ritual and a banquet. Chirac had abolished it in 1995, although the socialist Mitterand used the occasions for deal-making, including deals with ex-gaolbirds). The shoots that Sarkozy's fixer did succeed in arranging were a way of obtaining information for his master. Shooting remains a preoccupation of the international elite, of landed proprietors and of very large numbers of others, mostly men, even in developed countries — rising to 5.8% of the population in Finland, which is otherwise a paragon among educated nations.[iv]

Finland has a low density of human population and ample space in which to shoot. But in crowded countries, gunners are numerous too, 1.6 million in France for instance, where the political party called Shooting, Fishing, Nature and Traditions has both right-wing and left-wing support and actually secured six seats in the European Parliament in 1999. Shoots may be casually organised 'pot-hunting' but most nowadays follow stylised rules. Erich Honecker used to apologise profusely if he arrived at a shoot three minutes late and would never accept his keeper's advice to refrain from killing young deer.[v] What is to be seen in all this are the hand-me-down traditions of mass kills formerly organised for kings or emperors, like the Habsburgs. These traditions reached Victorian England in the form of the pheasant battue and they persist today.

At the battue, shooters wait while beaters drive the prey towards them. In continental Europe, the toll of game was huge, since prowess was judged by size of bag. The modern battue on English farms and estates is a variant, with young birds released in numbers at the start of the season. Individual gunners may or may not be adept at bringing down birds sent flying over them, though they do not do this with skilfully aimed shots but with a spreading pattern of lead from a 12-bore shotgun. What they no longer need is knowledge of the terrain, such as would come from stalking for themselves, as even rich men did until the early nineteenth century. Lavishing big sums on equipment and training means that some practitioners do become skilled, but knowledge of the countryside is not especially needed. Although its layout and ecology are strongly marked by sporting history, as far as modern sportsmen are concerned the countryside is little more than a painted backdrop.

For historians, too, the subject is at most a footnote. This blindness to a major influence on biodiversity in England goes back a long way. Foakes Jackson devoted one of a lecture series at Boston in 1916 to 'Sport and Rural England,' because he thought English life could not be understood without it, but he feared his audience would think the topic frivolous. Probably he was right. The subject continues to be skirted in rural history. As I have found from numerous conversations with historians and social scientists in several countries, the urban world has increasing difficulty in understanding agriculture, ecology or the countryside. Sportsmen reciprocate by caring little for the opinions of academe.

DUTCH SHOOTING ESTATES

The Low Countries offer a counterpoint to the English case with which this chapter will otherwise deal. The Dutch, urban as they early became, never stood aside from killing wildlife for pleasure: the vast medieval hall overlooking the market place in Haarlem was originally a hunting lodge. Taking part in blood sports was often the motive behind the establishment of country estates and, because of the interest of the House of Orange, hunting has been accepted as one of the costs of having a royal family.[vi]

Historically, landowners concentrated along the dune coast from The Hague to Zuid-Kennemerland, northwest of Haarlem. Court officials, followed by merchants from the maritime trades of Rotterdam, converted parts of the sand belt into estates, building large residences around the cores of existing farmsteads. These estates were less grand than those in England and, although each had its home farm, lacked the English surrounds of tenanted farms. This accentuates their recreational and social purposes, which in the eighteenth century featured fox hunting and shooting. The equivalents for Amsterdam merchants, who had once founded summer houses in the Beemster, were more purely residential, less single-mindedly sporting retreats built along the Vecht between that city and Utrecht.[vii] These were certainly houses in the country, though because they lacked subordinate farms or much surrounding land they did not have the connotation that 'country house' carried in England. From the sixteenth century, similar country places called *lusthof* or pleasure houses were built in Flanders by Antwerp merchants.

In Holland, as the urban economy developed, tycoons, beer barons and linen kings from Amsterdam and Haarlem erected houses on the edge of the dunes to escape the pollution of their own industries. The stench was worst in summer, and by the end of September, the women and children moved back to town, but the men kept returning until the end of October. Their visits were to hunt finches, which were thought a delicacy and were caught in thousands every season in fowling nets, using decoy birds. Trapping became an artisan occupation later; men of status hunted with guns instead.[viii]

The sporting appeal of the countryside is thrown into relief by some of the favourite subjects of Dutch painting. Merchants spent their wealth aping aristocratic lifestyles, including shooting, as was revealed in the artworks with which they adorned their houses. The shooting still life had been popular in Flanders and after 1650 became the vogue in the northern Netherlands. Paintings with game bags were especially admired. The quarry was not at that time reared game but whatever could be killed when rough-shooting. To choose a couple of examples at random, 'De provisiekamer' by David Teniers (1610–1690) in the Bredius House in The Hague depicts two bitterns hanging up, while a painting by Frans Snijders (1579–1657) in the Prinsenhof in Delft features a dead otter.[ix]

Since estates were intended as social statements, they fashioned and re-fashioned their settings by adopting successive fads in laying out and planting up their surroundings. Seventeenth-century pleasure grounds with formal French-style symmetrical flower beds (sport being restricted to the outer policies) were transformed into English-style parkland during the eighteenth and nineteenth centuries. Although little now remains from the French phase, the landscape to the north of The Hague is still held in estates which display arrangements of the tree planting prevalent in the nineteenth century.

This type of landscape can be seen to advantage in the Wassenaar district, which is an anatopism described as 'een stuk "Oost-Nederland" in de Randstad', that is, a displaced fragment of the wilder eastern Netherlands set down in built-up western Holland.[x] The royal estates are located here; hunting, latterly meaning shooting, took place on them and on the dunes from the Hook of Holland up through the Berkheide to Katwijk. From the sixteenth century, members of the House of Orange came here from The Hague to hunt. At De Horsten, early nineteenth-century stone posts can still be found, asserting the then Prins Frederic's right to hunt, while the Raphorst estate contains a tea-house that began life in 1863 as a lodge. It displays shooting trophies and the coats of arms of the jagtmeesters, masters of the royal hunt. The woods around are overgrown and, although studded with the old stones signifying the 'Jacht des Konings' (royal hunt), are penetrated today by public paths and are no longer a venue for shoots. The Jachtdepartement was closed in 1980.

Elsewhere in the Netherlands, shooting persists, though neither as extensively as in England nor with the same consequences for the shape of the countryside and access to the land. Dutch shooting arrangements are the subject of an insightful book by the anthropologist, Heidi Dahles.[xi] 'Hunting' (shooting), she writes, 'cannot be understood in terms in instrumental behaviour. When all the rational arguments in support of hunting have been adduced, its expressive functions remain the only convincing ones. Hunting indicates social status and serves male identity functions.' Among the Dutch, as among the English, it is largely a marker of social standing. For some of the urban middle classes — ironically labelled the 'status-hunters' by established participants — it is still an act of social climbing, partly because of its association with access to scarce land. 'Since access to land gains freedom, respect and power of life and death, hunting appeals especially' — this is Dahles — 'to the rich and powerful and to social climbers.' It is a world apart from mainstream Dutch society and although the number of licences issued has fallen and there is today an anti-hunting lobby, the participants form their own pressure group which is far from devoid of influence. That influence necessarily has some impact on the wildlife of the Netherlands.

SHOOTING IN ENGLAND

The international scope of blood sports means that speaking about the effects on landscape and wildlife is possible only in general terms. To impose some order on disparate experiences, I shall concentrate on lowland England, and on shooting, which, of the sports practised, was (and is) the chief influence on the environment. Since the eighteenth century, elements of what might be called a landscape of destruction have been superimposed on the productive landscape of agricultural England. As with the concept of a commodity landscape, the term is not proposed as descriptively complete but to draw attention to salient features. In this case, the countryside displays a meta-landscape whose commodities are birds managed purely for slaughter. Shooting was and is a consumption activity, a stylised pastime. The expense, the cost of harvesting the pheasant so to speak, has always exceeded the value of the game sold in butchers' shops and shot game are often simply discarded. In the past, profit was not the main concern of landowners who made shooting available to their guests as a matter of ritual and hospitality. Modern entrepreneurs who sell access to shoots are more concerned with rates of return.

Shooting has become big business. According to figures compiled by Peter Robinson, Fellow and past Principal of the Institute of Professional Investigators, the shooting industry was producing an annual return to estates of well over £ 1 billion pounds at the turn of this century.[xii] The effects on wildlife, the landscape and public access to the land are profound. Killing birds for pleasure interferes with farming and other productive activities to a greater extent than in other countries. Rural historians have unaccountably skipped over the topic.

The hypertrophy of what was originally a hobby and source of protein for landowners, and shooting for the pot or pest control among ordinary people, is surprising. May it not have been anticipated that an essentially medieval form of persecuting animals would either have dwindled in the course of time, like hawking, otter hunting or (finally) coursing hares, or — in a more humane and predominantly urban society — been legislated out of existence? What function can blood sports have that gives shooting and fly-fishing an enduring role and accords them exceptional political protection? One answer is the sheer scale of participation, a protection lacked by fox-hunting (and even there abolition barely succeeded). The number of people who shoot live quarry in the United Kingdom is given as 480,000.[xiii] This total is set to rise if 'shooting tourism' is expanded by the European Union's Rural Development Programme for England, which in 2010 granted the British Association for Shooting and Conservation (BASC) £ 1.6 million to develop game-shooting in Southwest England.[xiv]

No doubt individuals who are unconcerned about harming animals share a temperament that makes them indifferent to cruelty in general. As research into the issue concludes, 'how we treat animals... directly affects how we treat each other.'[xv] Nevertheless, with respect to blood sports, cruelty and death are merely collateral damage. For most participants, visiting pain and death on wild creatures may be discounted as an explicit goal; it is an intrinsic consequence but only a side effect. Interpersonal competitiveness is however central, as is shown by performance anxiety about the size of shooting bags, although maximising the bag is held in check by artificial codes of conduct. That this restraint irks some participants is apparent in advertisements for sporting tours of Eastern Europe, which lay heavy stress on the absence of bag limits: 'you decide when to stop shooting.'

Among traditional motives for shooting in England, Mark Girouard notes that during the 'Age of the Pheasant' (1865–1914), shooting parties in country houses were displacement activities for the younger sons of county families who had inherited incomes of only a few hundred pounds per year, too small to maintain a wife.[xvi] They spent their leisure travelling from house to house and shoot to shoot, engaging in this and other activities that young bachelors might be expected to find attractive.

The common thread is male bonding, though somewhat paradoxically social climbing is also implicated. The private character of shoots, not to mention their cost, kept out *hoi polloi.* Queen Victoria preferred fox-hunting as (in some sense) open to all, contrary to the social exclusiveness of her son's shooting parties. Veblen went to the heart of the matter when he commented about the conspicuous waste of time and resources, saying that without the social function, 'it is quite beyond the reach of imagination that any adult male citizen would of his own motion go in for the elaborate futilities of British shooting'.[xvii] As we saw, Dahles commented in the Netherlands that only the expressive functions of shooting make sense. Yet there is an important commercial element: business and political alliances are formed at shoots and the meals that follow them, and this extends to international deals now that rich people can fly in and out of the country to take part. For many participants, business is the true *raison d'etre.*

American writers have been quick to recognize the stylised nature of blood sports in the British Isles.[xviii] This they attribute to their origin in aristocratic pastimes, which is correct as far as it goes, although rather than being genuinely ancient the social form descends from the Victorian battue. The precise history is a secondary issue, involving tangled stories of improvements in gun technology, developments in estate ownership, the professionalization of game-keeping, changing methods of rearing game-birds and a never-ending campaign by the keepers to exterminate any likely, and many an unlikely, predator on game. Technological innovations have to be kept in check or there would be few birds left. Social behaviour has remained conservative. 'The cycle of the seasons and many traditions associated with our sport have not changed for centuries. It seems fitting, then, that the 9 September 1882 issue of *Shooting Times* bears an uncanny resemblance to today's.'[xix]

The highest score for head of game killed was a matter of contest. Keepers were required to put on a good show of birds for the sake of their employers' prestige. Nevertheless, they wanted to keep something in reserve and all-out slaughter was frowned on to the extent that some species of animals and birds were never intentionally killed. Foxes were not shot — fox-hunters regarded this as more than a crime, a sin. At the appropriate season, hen pheasants were not to be shot, only the cocks, and so forth. It is possible to interpret such restrictions in two ways. As with most sports, the rules were set so as to make engagement reasonably difficult without being impossible. The prohibitions introduced piquancy. Some Americans, accustomed to a more lavish environment and the lesser restraints it placed on the gunner, found this baffling. Eventually, they began to suspect a second rationale in a form of conservation, if only to safeguard the estate's own stock of game; restraint does not extend to so-called vermin. The overall effect of game preservation is to distort ecosystems.

ALTERING THE ENVIRONMENT

Each time cereal prices fell far enough to cause bankruptcies among farmers, market town shopkeepers, lawyers and even schoolmasters found they could afford to shoot. They formed syndicates to rent the shooting from temporarily embarrassed landowners. In the worst times for farming, sporting rents exceeded agricultural rents. Farmers then complained they could not afford to shoot over their own land. There was a double trap, with shooting monopolised by the owners and occupiers of land in good times and the vacuum of bad times filled by *ad hoc* syndicates. Either way the shooting continued.

Shooting practices were remoulded over the centuries, and any brief description must make the phases seem more distinct than they really were. The very structure of vegetation began to be altered to suit the sport. Whereas in the rough countryside of the early eighteenth century, sportsmen had found it hard to get clear of blackthorns, by the 1780s rides were being cut through estate woodland. At the same time, rough hedgerows to harbour game were actually being encouraged. Tenant farmers might justly object that hedges soon encroached on the fields, but their voices counted for less and less when shooting became so fashionable.

In general, the effect lent towards taming the countryside, though leaving it in a mildly arrested stage of development compared with what maximum agricultural and timber production might have created. During the late eighteenth century, plenty of landowners became keen on improving their farms but sporting fashion was so powerful that even the improvers were inclined to give precedence to sport. The tide of cropping intensification and landscape innovation had the result of developing habitats for game: more turnips and wheat fields meant more pheasants and partridges, while more ornamental spinneys also favoured pheasants. But spinneys favoured wood pigeons too: if you farm pheasants, you will farm pigeons.

Wood pigeon and pheasant both like wood margins for nesting and roosting but need ready access to open country for feeding. To accommodate these tastes on the part of the pheasant, and inadvertently on the part of the pigeon, the practice was to plant small woods that have long perimeters relative to their area. Where larger stretches of woodland were planted, they also tended to be made as long and thin as possible. They were wider than shelterbelts, but not much wider, and besides harbouring pheasants and foxes served to protect crops from wind and weather. Many of these linear woods still wind up hill and down dale for miles across the chalk downs and limestone wolds. They are silent testimony to the fact that what strikes the eye as plain farmland doubles as sporting country. It was meant to do so. The most renowned agricultural estate in the realm, Holkham, Norfolk, was laid out for the shooting.

Remodelling estates by means of tree-planting became commonplace, as for instance (to cite a single example) at Micheldever, Hampshire, in the late eighteenth and early nineteenth centuries.[xx] A century later, about 1935, when J. Arthur Rank bought the 10,000 acres of Sutton Scotney manor only three or four miles from Micheldever, he too planted belts of fir, spruce and larch, with privet and Marabella plum on either side. He also established two or three remises (feeding places) planted with grain.[xxi] During the 1930s, industrial capital was entering shooting estates and names like Reckitt and Fairey joined or replaced those of longer-established landowners. The trend continues, with rich foreigners prominent among purchasers during the first decade of the twenty-first century, though in southern and central England, it is City financiers who tend to be the buyers. It makes little difference to estate management.

Blood sports were, and are, a major motive, sometimes the chief motive, for acquiring an estate. During the Age of the Pheasant, few would have found the subordination of farming to the gun unusual. When it came to where and what fodder crops to sow, farmers bowed to the dictates of their head shepherds, but the head keeper (who tended to have the landowner's ear) trumped even the shepherd, as well as dictating to the estate woodmen where new planting should take place.[xxii] Tenant farming, though important, took second place — and by Victorian times, the large majority of English farmers had become tenants. Admittedly, there are examples where the influence ran the other way and the shoots adapted to changes in methods of farming. One instance was the shift to employing beaters and shooting driven birds once partridges took to kale fields so wet the gentry did not wish to wade through them. It might therefore be said that sport and agriculture co-evolved.

CONSEQUENCES FOR WILDLIFE

In his essay 'Nature and the Gamekeeper', Richard Jefferies describes the essentials of the impact of Victorian game preservation on wildlife, although he was too good a naturalist to lay every change at the door of the pheasant coop.[xxiii] He points out that some of the pressure exerted on predators by the gamekeeper was not intended solely to preserve game but to protect rabbits, the meat and skins of which often paid his wages. The wide rides cut for pheasant shooting favoured the rabbit (and, equally inadvertently, light-tolerant butterflies).

Of 20 species of birds and animals pressed hard by the Victorian keeper, Jefferies totals up 9 that had become scarce or completely exterminated in lowland England by the third quarter of the nineteenth century. The list included the wiping out of pine martens and polecats and extended to the killing of barn owls and tawny owls. Another 11 species, including stoats, weasels, jays, magpies, kestrels and (surprisingly) sparrow-hawks, had survived despite everything keepers could throw at them. Another surprise over which Jefferies muses was the carrion crow, hated by everybody in the countryside but persisting nevertheless, whereas the larger corvid, the raven, had been driven to extinction. Additionally, he points out, squirrels were favoured by the woodland habitat created in order to rear pheasants. He meant the red squirrel, which was actually on the increase at the end of his life in the 1880s, before succumbing to epidemic disease at the start of the twentieth century.

Jefferies' expertise was gained by first-hand observation in Wiltshire, chiefly close around Coate, near Swindon, in the woods of the Burderop estate. In Hampshire, we have similar testimony from Col. Richard Meinertzhagen, who lived as a boy from 1884 to 1900 on the River Test at Mottisfont. Meinertzhagen, who knew about such things since his own family employed keepers, told an even more scarifying tale than Jefferies.[xxiv] He reported that, 'no kestrel or owl or hedgehog was ever allowed to live; keepers would loaf about with a gun shooting everything big, including woodpeckers. Many estates had exterminated hawks, owls, rooks, jackdaws, herons, little grebe, moorhen, coot, water-voles, badgers, squirrel, and hedgehogs under the impression that game alone should be allowed to live.'

In further corroboration, W. H. Hudson wrote in *A Shepherd's Life* (1910) that 'the curse of the pheasant is on... all the woods and forests in Wiltshire, and all wild life considered injurious to the semi-domestic bird... all the wild life that is only beautiful, or which delights us because of its wildness, from the squirrel to the roe-deer, must be included in the slaughter'.[xxv] Hudson was more sweeping in his condemnation of game preservation than Jefferies. The latter's realistic appraisal of the effects add up to a mixture of the Allee effect, which led to local or regional extinctions when individuals were too few to find mates, and 'rat-farming', which meant that 'vermin' quickly recovered through recolonizing the best sites for nesting and feeding. The conclusion is that in the years before the First World War many species of birds and mammals were hard pressed, but few were more than regionally exterminated. Even in the face of the pressure from the shooting interest, there seems no reason to believe in a downward trajectory for populations of wildlife from which there could be no reversal. Most would eventually recover.

THE MODERN SITUATION

The number of gamekeepers is far lower today than at the peak in 1914. Specialist rearing farms from which the stock of young pheasants can be continually replenished have removed some of the incentive to kill 'pests'. The public frowns on the indiscriminate killing of potential predators (though oddly not on the industrial-scale slaughter of game-birds) and the gibbets hung with the corpses of 'vermin', which adorned every wood edge a few decades ago, have disappeared. The shooting industry is larger than ever but as far as one can tell the killing of vermin is not as relentless as it once was. The catch lies in the phrase 'as far as one can tell', because access to the countryside is controlled far more strictly than would be needed to protect farm crops. Any activity may be taking place unregulated and out of sight.

Propaganda by the shooting industry insists that its adherents do a great deal of unpaid conservation work. This is disingenuous. Game preservation implies landscapes where the gamebird is a commodity additional to farm products. It may make the countryside more varied than arable prairie, but is scarcely optimal from the public interest or environmental points of view. What may or may not be done on land to which there is no entry is of limited social value. There is interference with public footpaths. Woodland cover and crops remain influenced by the dictates of shooting. Maize strips planted to 'hold' gamebirds are magnets for badgers and deer, which are harmful elsewhere. Releasing an estimated 20 million pheasants into the countryside every year has serious implications. The mistake is to believe they live exclusively on food put down by the keepers, on spilled grain, and weed seeds. Pheasants are omnivorous and work the hedgerows hoovering up every edible it is possible to conceive. They are 'ecologically disastrous', although this is rarely mentioned in discussions of threats to wildlife.[xxvi]

Further negative spill-overs from shooting are various, including the damage flying birds do to electric power lines and the danger to traffic from the semi-tame birds straying onto the roads like chicken. The costs do not fall on those who release them. At a more abstract level, a distortion of the whole of English society is admitted by Max Hastings, himself an enthusiastic shooter.[xxvii] He observes that the quickest way for a rich tycoon to enter society is to buy a grouse moor [or pheasant shoot] rather than endow a hospital or art collection. Hastings emphasises the social attractions of shooting and claims that, 'the traditional sporting enthusiasms of the aristocracy strongly influence the manner in which social ambitions are pursued.'

Moreover, shooters, including wildfowlers, have succeeded in holding onto the right to kill species that are neither game nor wild birds protected by law. These 'grey' species are primarily waders such as golden plover and woodcock, which are not reared for sport but simply appropriated as quarry. They are wild species that ought to be available for everyone to enjoy. Woodcock, in particular, are tracked across Britain on migration and targeted by a specialist band of hobby gunners. The public, including, of course, the naturalist, has an interest in the preservation of these birds but is obliged to accept that, although they are not game, they may still be shot. The justification seems to be nothing more than inertia. There is no check on how many are destroyed and in the sporting press there is much boasting about woodcock bags, the size of which is regulated only by the shooters themselves.

An argument that the birds which are killed are surplus to the breeding stock would be irrelevant; what is at issue is the opportunity for members of the public to watch and enjoy all of them while they are alive. The contempt in which the public interest is held in this respect is captured by a statement at a public hearing in 2002 by an adviser to the National Gamekeepers' Organisation (and formerly of the Game Conservancy Trust).[xxviii] He declared that, 'provided bird-watchers can come and actually see a Golden Plover or whatever (sic), you may not necessarily need a huge number of Golden Plovers to satisfy the objective of the reserve.' This would legitimise the diversion into private hands of what is rightly public property, as well as ignoring the sublime experience of watching big flocks on the wing. Shooting thus presents a social problem; given a different balance of political power the environmental distortions it involves could be rectified by amendments to law and public policy.

ENDNOTES

[i] Review of L. Cassels, *The Struggle for the Ottoman Empire, 1717–1740* (London: John Murray, 1967), in *The Times Literary Supplement*, 17 August 1967.

[ii] J. L. Anderson, personal communication.

[iii] *The Times* 13 February 2010.

[iv] *The Economist* 11 September 1999.

[v] *Daily Telegraph*, 6 September 2008.

[vi] R. J. E. M. van Zinnicq Bergmann, *In dienst van drie vertstinnen: Het leven van een hofmaarshalk* (Heerenveen: Van Mazijk, 1998). The author was Queen Juliana's Jagermeester.

[vii] *Ruimsteijke kwaliteit kastelen en historiche buitenplaatsen/Zuid-Holland* (OKRA Landschapsarchitecten: Stichting PHB, 2007), p. 11.

[viii] See the exhaustive study by I. Mathey, *Vincken moeten vincken locken: Vijf eeuwen vangst van zangvogels en kwartels in Holland* (Hilversum: Uitgeverij Verloven, 2002) and H. Felling, *Buitenplaatsen in en om Den Haage* (Zwolle: Waanders, 1992), pp. 30–33. I am indebted to Mr. Albert Niphuis for these references.

[ix] Hunting still-lifes are discussed in the catalogue of the Kremer Collection, a touring exhibition seen at the Frans Hals museum, Haarlem, 2009.

[x] F. Steenkamp and N. van Gelderen, *Natuurlijk Wassenaar* (Wassenaar: Vrienden van Wassenaar, 1992), p. 55.

[xi] H. Dahles, *Mannen in Het Groen: De wereld van de jacht in Nederland* (Nymegen: SUN, 1990), pp. 3–7, English summary, 'Men Dressed in Green: The world of hunting in the Netherlands'.

[xii] P. Robinson, Pheasant Shooting in Britain: The Sport and the Industry in the 21st Century. Available at: www.animalaid.org.uk.

[xiii] *Shooting Times*, 9 December 2009.

[xiv] *Shooting Times*, 21 July 2010.

[xv] G. Hodson and K. Costello, 'The human cost of devaluing animals,' *New Scientist*, 15 December 2012.

[xvi] M. Girouard, *Life in the English Country House* (Harmondsworth, Middlesex: Penguin, 1980), pp. 295–296.

[xvii] T. Veblen, *Imperial Germany and the Industrial Revolution* (London, 1995 [originally published in 1915]), p. 142. See also E. L. Jones, 'The Environmental Effects of Blood Sports in Lowland England since 1750,' *Rural History*, **20** (1) (2009), pp. 51–66.

[xviii] See e.g. J. Graves, *Goodbye to a River* (New York: Vintage Books, 2002), pp. 244–245; W. Humphrey, *Open Season* (New York: Dell, 1989), pp. 196–197.

[xix] *Sporting Times* 12 September 2012.

[xx] A. B. Milner, *History of Micheldever* (Paris: privately printed, 1924), p. 37.

[xxi] Anon. (ed.), *British Sports and Sportsmen: Shooting and Deerstalking* (London: Sports and Sportsmen Ltd, n.d. [c.1938] p. 202.

[xxii] An example of a head keeper's power over estate affairs is given in Deborah Devonshire, *Home to Roost* (London: John Murray, 2009), p. 14.

[xxiii] R. Jefferies, *The Life of the Fields* (Oxford: Oxford University Press, 1983) [first published 1884], pp. 65–69.

[xxiv] M. Cocker, *Richard Meinertzhagen* (London: Mandarin, 1990), pp. 27–28.

[xxv] W. H. Hudson, *A Shepherd's Life* (London: J. M. Dent, 1936), p. 210.

[xxvi] H. Warwick, *The Beauty in the Beast* (London: Simon and Schuster, 2012), pp. 291–292.

[xxvii] *Financial Times* 13 November 2011.

[xxviii] Available at: www.defra.gov.uk.

7. LANDSCAPES OF DESTRUCTION: THE SACRIFICE TO TROUT

The man of creative imagination may find his consolation in the flash of kingfisher but for most of us fishing is the thing.

Howard Marshall, *Reflections on a River* (1967)

CHALK STREAM ANGLING

Like shooting estates, dry-fly venues on the chalk streams are rarely accessible to the ordinary citizen. They have been converted into playgrounds for the extremely rich, whose inspiration may owe less to the beauty of preserved waters than to satisfaction with their own privilege. As Howard Marshall wrote, 'I do not suppose that the stockbroker, escaping from the city at the weekend, drives eighty miles and dresses up in thigh boots and a fishing jacket, and a most peculiar waterproof hat, in order to hear the drumming of the snipe'.[i]

The insensitive stockbroker is heading eighty miles from London to fish on the streams of the south country. These, above all the Kennet in West Berkshire and the Test in Hampshire, are the ones on which I shall concentrate, in order, once again, to restrict the array of examples. At least the venues are superb. Ironic though it may be that a countryside devoted to pursuing wild creatures should be thought especially attractive, this is the case. It includes the chalk stream valleys that are the setting for dry-fly fishing, the form of angling with the highest status. It would be unfair to think that the scenic attraction goes unnoticed by anglers; not all are insensitive stockbrokers. Angling books may start with enthusiastic sagas of catching fish but not infrequently tire of recounting the score and end with paeans of praise to the *mise en scene*.

Writings about fishing, of which there are even more than about golf, are to an extraordinary degree taken up with the arcane lore of dry-fly fishing. They record details of equipment and method until the reader's head spins, they list the weights of the catch *ad infinitum* and describe disputes so vitriolic they scorch the paper — whether to cast upstream or downstream or whether to adhere to wet fly or dry fly. Fishing brings out the worst in its proponents. Even catch-and-release, where the fish are not deliberately killed, amounts to toying with live creatures as playthings. Having a fish on the line brings out the 'potential psychopathy of power wielded from a distance'.[ii]

Reputations were made or ruined in the nineteenth century according to the side one took in the disputes. Two great protagonists at the end of that century, Halford and Skues, contested mightily over how to catch the brown trout. Tony Hayter, who chronicled their battles in *F. M. Halford and the Dry-Fly Revolution*, quotes a review which understandably noted that, 'Mr Halford is rather too much inclined to make a business of what should after all be a recreation'.[iii] The saving grace, if such it be, comes from the insight of the Texan, William Humphrey, that the latent function of all the stylised practices is to share out scarce resources.[iv] The resources are not, however, shared with the common man.

An unfortunate consequence of the obsession of anglers with technical minutiae is that, in England, they have written little serious ecological history. There are only sparse comments from which to piece together the ways that fishing affects riparian habitats and the other wildlife living there. Initially, of course, the streams were fished by men who lived nearby and intended to eat their catch. Mostly they wielded nets but were prepared to use whatever device worked — in 1658 it was decreed that any commoner at Hungerford on the Kennet might by custom fish, 'with hooke and lyne and with rodden and with his hands'.[v] In other words, the age-old country practice of tickling trout was officially sanctioned. Hungerford was special because its householders clung tenaciously to their common ownership of the fishery. The property was worth money to them, and remains so, for since early modern times topographers have noted that the Kennet at Hungerford is famous for trout and crayfish. To safeguard their rental income, the commoners accepted restrictions on their own fishing — limits on the days they might fish and the tackle and bait they might use.[vi]

The bulk of riverside communities were less lucky and their fishing was privatised in the eighteenth and nineteenth centuries. As with enclosure of the land, enclosure of the rivers and river banks was hard for poor and dependent villagers to resist. Landowners found it worthwhile to acquire the rights, partly because freshwater fish could be sold on the London market: for instance, fish were trapped and despatched to the city from the Littlecote estate, near Hungerford. Even before railways made the chalk streams more accessible to Londoners, sport was however taking over from commercial fishing. On the river Test, the secretary of the Houghton Club was using a phenological technique in 1820 to calculate the day in April when the grannom fly would rise at Stockbridge and when the mayfly would rise at the end of June.[vii] He watched the sequence of natural events, which is what phenology means. London was rural then and he could note the blooming of whitethorn, elder and guelder rose, and the coming into seed of common hedge garlic. At that point, the fly would rise and the trout would follow. Fishermen planned to follow too, taking the coach to arrive on the river at the precise moment.

When rail opened up the south country, the number of rich Londoners descending on the streams grew enormously and the price of fishing rose accordingly. Hatcheries were established and restocking took place. Landowners found they had at hand a slightly unexpected source of income — at least they had not anticipated its scale. They built fishing lodges to let, just as they built shooting lodges. When farm rents fell during the depression of the late nineteenth century, this lucrative alternative grew fast. Occasionally, local inhabitants tried to assert or re-assert their right to fish, but lost out. The last, dying fall was at Chilbolton on the Test, where a series of costly law suits stretched into the twentieth century but ended with the Church of England prevailing as piscatorial owner.

The chalk streams became ever more pricey; they have ended as upmarket fish farms. Those who own or manage them believe this to be part of the natural order and form such a powerful elite group they have been able to resist every intrusion. During the Second World War, scientists experimented in trying to set pontoons alight by sending burning fuel down the Test, but even during a national emergency 'were forced to abandon their tests because of the effect on fish stocks'.[viii] This was not because the fish were being farmed as part of the food supply! Some modern works on fishing deny every other interest in rivers and river life. They make an ambit claim for the right of sport fishermen to 'discourage' or kill whatever seems to threaten trout or their fry: mute swans, little grebes, grey wagtails, and so forth.[ix] The assertion of privilege extends to leaving the river banks too rough for families to picnic.

RIVER MANAGEMENT

Little is known in detail about the historical re-engineering of the chalk valleys for the sake of fishing, very little indeed about the dates at which changes were made. It does seem that until the seventeenth century, sometimes later, the valley bottoms were boggy, full of rough herbage and wandering rivulets. The pasture was grazing for village livestock and as late as the start of the nineteenth century weed from the rivers was used in places as fodder for cattle.[x]

By the time in the seventeenth century, when Isaac Walton went angling on chalk streams, modifications had been made to the water regime in order to 'float' the meadows. This involved hatches, drains and ridge-and-furrow, whereby the water was directed over the grass to bring on early growth, and then back into the main channel. It was a boon to have grass for the sheep during the 'hungry gap' of April, when stocks of fodder had been used up over the winter.

An exceptional set of materials on river and fishery management survives in the annual accounts of the Constables of Hungerford between the mid-seventeenth century and the late eighteenth century, and in reports of commissions of enquiry into the water management there over a longer span.[xi] Payments were made for cutting weeds, cleansing the river of mud and wood, putting hurdles and stakes in the river, keeping cattle out, dragging the river (apparently with a horse rake), and throwing gravel out of the bed. There were receipts for permitting water to be diverted for floating the adjacent meadows and for the annual rents of fishing. The rents were paid by only two or three individuals, presumably men whose livelihood was catching fish for sale. Significant use by sport fishermen, other perhaps than Hungerford commoners themselves, seems to have awaited later periods.

The borough's ownership of the fishing rights instigated plenty of enquiries and disputes. Three-way conflicts between millers, farmers who wanted water for irrigation and rivers keepers concerned with the fishing, were part of the history of this and every chalk stream. Hatches were always being surreptitiously drawn or closed by parties who wished to steal a march on the others. At Hungerford, the fishing rights generated frequent claims and counter-claims, and at intervals commissions were appointed by the Duchy of Lancaster to look into them.[xii] The issues included rights to fish, the correct boundaries, episodes of poaching and whether there was enough water for all the mills. It was not all earnest labour: one witness said of men who fished together, that 'when they took a good trowte, they would make a great noyse and shout for ioye.'

In 1608, there was a dispute as to whether a mill that had been rebuilt at Oakhill, above Hungerford, was interrupting the flow of the Dun, a stream that joins the Kennet near the town. The commissioners were extremely thorough, visiting all the mills and streams round about, besides undertaking an actual experiment: they shut up the Oakhill flood-gates 'and sett up an howre glasse' to time how long it would take to block the flow (only one hour). They can scarcely have known of contemporary Italian work on hydrodynamics, which was not translated until after the Restoration, but their approach was more scientific than ordinary English practice.[xiii]

All this detail shows how valuable the rivers were and how many interests jostled to use them, though it does not provide a really comprehensive view of management or indicate how hard pressed the fish and wildlife were. It is difficult to be sure what happened at given periods. Isaac Walton, who wrote fishing's Bible, may have fished the little side streams in the Itchen water meadows rather than along the banks of the main rivers. Only in the eighteenth century were river banks cleared of enough trees to make casting easy. Meanwhile floating the meadows was spreading at a scarcely known rate and some schemes may merely have revamped earlier works. Irrigation may have been regularized in the late eighteenth and nineteenth centuries to carry greater stocking densities of livestock. While art is not necessarily a reliable witness, prints and paintings from the early nineteenth century do not yet show the braided channels subsequently characteristic of the valleys. It was in the mid-nineteenth century, with an upturn in the angling interest, that management of the rivers really did intensify. Weed on the bottoms was then mown with the scythe three times per year, to stop the rivers becoming choked and clogging the mills. Owners of fisheries, too, were annoyed by the mud and clumps of weed moving down on them and eventually the times of clearings were co-ordinated.

Yet nothing at mid-century matched the effort that was to go into river improvement during the final decades of Victoria's reign, when more hatcheries were established and there was much restocking. Fishing with the dry-fly had become a craze, very profitable to landowners. The streams were ever more closely tended, cleared of weed more often, and subjected to blitzes to remove pike and coarse fish. The keepers fought a perpetual but inconclusive battle with coarse fish, especially chub, because they competed with the trout. Photographs exist of row upon row of dead specimens to show the scale on which coarse fish were removed. Hayter reports that a trout fishery on the river Wylye in Wiltshire had been allowed to fall derelict until a programme of renewal was instigated between 1891 and 1895.[xiv] Pike, 3,619 of them, were killed in those years, together with 13,056 other coarse fish — think of the effort that went into the counting alone! But labour was cheap.

An important point emerges from this, which is that the incessant labour involved in river management tended to produce only semi-transient effects. Coarse fish might be removed but proved impossible to eradicate completely or prevent from recolonizing the waters. The environmental impact was therefore cyclic rather than permanently downwards. Draining the valley bottoms may seem to have made a lasting mark; the side ditches, drains and ridge-and-furrow remain as archaeological evidence of former floated meadows. But from the hydrological point of view they retain only minor significance.

The late nineteenth-century depression dealt a blow to the economy of folded sheep, water meadow lost its value, and the hatches were soon crumbling and dilapidated — like everything else in agricultural England, wrote Alfred Williams in 1912.[xv] He recorded a penny of 1792 found at a hatch on the Cole, near Swindon, which may indicate the date of the last new work. Perhaps the time will return, Williams speculated, when irrigation of the meadows will cease, 'except for what is performed naturally.' As it was, even in its twilight phase, 'the whole course of the river is flooded half a mile wide; it is like an inland sea. Enormous crops of grass follow these inundations — the yield is often three tons to the acre.' Dramatic though this may have appeared, it was not inconsistent with a rich population of fish. The pools below the hatches teemed with roach, dace, pike, sometimes trout, less often perch.

USES LIKE NESTS OF BOXES

River valleys thus had many uses alongside fishing. At different times, often the same time, they were the sites of grazing, of mills for fulling, cloth, corn or paper, and of peat digging, gravel digging, osier growing and tanning. Mills were sited at every nick point where there was a fall to drive the machinery — the chalk streams offer only slight falls, but when transport was dear and grain was grown in every parish to save the cost of carting it far, no good site was left unoccupied. Mills are venerated by painters and the public as deeply historical buildings but in truth they have no stable history. The sites were geomorphologically fixed by the fall at natural nick points in the stream but the structures have repeatedly been rebuilt; it is only their twilight form which is to be seen today. They shifted between various purposes, ending with the lowly occupation of grinding grist for animal feed, until in the twentieth century all active use gradually ceased and the mill houses were left as dwellings.

The main use of the valley floor was as pasture, irrigated where possible: according to a list of 1808, almost 1,400 acres were irrigated between Hungerford and Reading.[xvi] An additional 230 acres were being used for digging peat, which was burned and its ash used as fertiliser. This was a local speciality. Peat had been burned to heat cottages but the fertiliser industry took over from the early eighteenth century, when it was realised the ash encouraged clover. The procedure lasted until about 1870. The manufactory near Newbury where the peat was burned was thought distastefully industrial but the diggings themselves, once they were worked out, were soon assimilated to the rural scene.[xvii] The excavated land was converted into water meadow, for example, at Hoe Benham on the Hungerford side of Newbury in the early eighteenth century.[xviii] The only interference with the peat industry there was a piece of bullying by Lord Craven, who was busy floating his meadows. A tenant dug a ditch to drain his excavations and Craven had it filled in, claiming water from the peat killed his fish.

In the 1730s, 'Dutch willows' were planted on the land that had been dug for peat at Hoe Benham. This marked the inception of the valley scenery to be seen today: water meadows were formed on the peat diggings and planted around with pollarded willows, as well as with the osier beds that are commonly mentioned. An account of early nineteenth-century peat digging and osier production comes in the papers of George Frankum, a yeoman farmer at Woolhampton.[xix] Osiers, grown for the very extensive (and much neglected) trade of basket weaving, were planted on the wetter meadows and harvested into the twentieth century.

Peat-burning may have been unique but in other respects, the Kennet was typical of chalk streams. All the valleys now lie with the signs of sequent occupance upon them, the most visible modern addition being wet gravel pits. The earliest pits were dug on the terraces at the sides of the valleys, but about 1960, when pumps were introduced, excavation moved to the valley floor. Pumping kept the pits dry enough for working but when the pumps were switched off, the pits filled with water, attracting an array of modern users to match the variety of the more rustic past: fishermen, birdwatchers, developers of second homes, dinghy sailors, and water ski enthusiasts. The valleys are crowded because the modern population is larger and richer. But the needs of past centuries had also meant intensive use.

'THE SACRIFICE TO TROUT'

In 1884, in his book of essays called *The Life of the Fields*, Jefferies wrote a short piece called 'The Sacrifice to Trout', which will ring true to any naturalist.[xx] He listed 16 creatures that were being thinned out across the south of England as a result of preserving trout. The otter came closest to complete extirpation at the hands of the river keepers; it is only now recovering. Among fish, pike and perch were systematically killed. Among birds, heron, kingfisher, moorhen, coot, mute swan, mallard and teal were all persecuted. Herons were wary and hard to hit with a shotgun, but small bore rifles came along that could kill them at a distance. The nests of moorhen and coot, duck and swan were destroyed — birds already reduced in number by the tidying of river margins and reduction of marshy spots. The species mentioned were the keepers' main targets. The charge was that they were competitors which ate trout fry. William Lunn, the Houghton Club keeper from 1887, killed 150 little grebes per year.[xxi]

FISHING: IN CONCLUSION

Wildlife in the south of England was hit hard by the craze for blood sports because there was so little untended land to act as a refuge. Unlike much of the north of the country, the streams ran through cultivated fields and there were few places to which wildlife could escape. From mid-Victorian times, the river banks were half cleared of trees to make room for casting flies, though some trees were left to give the shade that trout like. At least this was an improvement compared with early Victorian times, when riverside trees had been lopped or felled wholesale because the blowing line was in use. Uninterrupted breezes were needed to carry the line across the river. The chalk valleys must have looked more open during the early nineteenth century than they have done since and may have been less productive of insect life.

Chalk-stream valleys form palimpsests of superimposed uses; they may be said to be scarred by the tumble of past activity, if scarring is a reasonable term. But surprisingly it is not. The ensemble creates an environment which by common repute is thought attractive today. Each of the components — abandoned water meadow, overgrown osier beds, wet pits — forms its own ecosystem, and each is a magnet for naturalists. Exploitation has introduced a variety that was not present previously.

The argument advanced by fishermen, that the conservation of fish is good for other wildlife, is self-justifying. Everything depends on what mix of wildlife is to be preferred, and by whom.[xxii] Some species continue to be suppressed by culling, and what amounts to the gardening of vegetation takes place in the streams and on the banks. Although centuries of effort have not succeeded in bringing about the permanent extinction of the trout's enemies, this is beside the point. The public interest is spurned by the very attempt. Even were the benefits of fishery management unequivocal, a narrow group still monopolises riparian rights, chooses in its own interest what ecosystem may flourish, and excludes the general public from enjoying it.

Yet despite the distortions by way of managing habitats and engrossing property rights, there is no good case for supposing that the riverine environment itself has deteriorated step by step. At intervals, it has been drastically altered but not necessarily for the worse and not irrevocably. Its history has not been linear; rather the uses of the river valleys have been added to or replaced in response to changing demand. What we observe is a long and varied sequence of ecological changes chosen by the few instead of the many. Secure private ownership is the foundation of economic life but it is hard to believe that on the chalk streams it was always acquired fairly or that it does not create serious negative externalities.

ENDNOTES

[i] Quoted by C. Voss Bark, *The Encyclopaedia of Fly Fishing* (London: Batsford, 1986), p. 94.

[ii] *Financial Times Magazine*, 22 May 2010.

[iii] T. Hayter, *F. M. Halford and the Dry-Fly Revolution* (London: Robert Hale, 2002).

[iv] Humphrey, *Open Season*, pp. 196–197.

[v] E. L. Davis, *The Story of an Ancient Fishery* (Hungerford: Trustees of the Town and Manor, 1978), p. 18.

[vi] F. Thacker, *Kennet Country* (Oxford: Blackwell, 1932), p. 138.

[vii] R. Deakin, *Wildwood* (London: Hamish Hamilton, 2007), p. 23.

[viii] M. Gillies, *Waiting for Hitler* (London: Hodder and Stoughton, 2007), p. 201.

[ix] See e.g. B. Clarke, *On Fishing* (London: Collins, 2008) or M. Lunn, *A Particular Lunn: One Hundred Glorious Years on the Test* (London: A. & C. Black, 1991).

[x] Rev. Willis, 'On Cows for Cottagers,' *Annals of Agriculture*, **40** (1803), pp. 55, 564–567.

[xi] Transcripts of the accounts of the Constable of Hungerford re river and fishing, and law suits relating to the Duchy of Lancaster, Hungerford Virtual Museum, Internet version.

[xii] Davis, *Ancient Fishery*, p. 12, notes that on 24 April 1599, the claims were sworn before Henry Masters and William Jones, Commissioners. Masters was probably an ancestor of the Chester-Masters family of Cirencester and Jones possibly an ancestor of mine from Woodlands, Mildenhall.

[xiii] T. S. Willan, *River Navigation in England 1600–1750* (New York: Augustus M. Kelley, 1965), p. 79.

[xiv] Hayter, *Halford and the Dry-Fly*, p. 166.

[xv] A. Williams, *A Wiltshire Village* (London: Duckworth, 1912), pp. 24ff.

[xvi] W. Mavor, *General View of the Agriculture of Berkshire* (London, 1809), pp. 538–539.

[xvii] H. J. E. Peake, 'The Origin of the Kennet Peat,' *Transactions* of the Newbury District Field Club VII/2 (1935), pp. 116–126. Peake dates the origin of the fertiliser industry too late.

[xviii] Papers relating to Hoe Benham, Berkshire Record Office D/EAh E2 (1731–1734); D/EAh: E15/2 (1638–1830).

[xix] C. Knight, *The Life and Times of George Frankum 1763–1830* (Winchester: Knight Publications, 2000).

[xx] Richard Jefferies, *Life of the Fields* (Oxford: Oxford University Press, 1983) [First published 1884], pp. 170–174.

[xxi] Lunn, *A Particular Lunn*, p. 121.

[xxii] The military has even canvassed filling in all Upper Thames gravel pits in order to reduce the number of Canada geese that threaten aircraft.

8. THE NETHERLANDS: RECLAMATION AND EXPLOITATION

> One cannot resist speculating that the Dutch rural scene that is now such a source of pleasure and contentment — a pasture landscape with windmills looming in the distance and a lake in the foreground — must have been a source of anxiety to the observer of 1500.
>
> Jan de Vries and Ad van der Woude (1997)

PROBLEMS OF AN EMERGENT LANDSCAPE

The Netherlands, especially the province of Holland, were and are anomalous, much of the surface having been rescued from estuarine mud and sand or reclaimed from marsh and peat bog.[i] A large proportion lies more than one metre below sea-level and would flood if there were no banks to protect it.[ii] Here the Dutch very early created an economy out of all proportion to their land's small size. This is the world's most artificial country and it is no surprise that its wildlife communities are constructs of human land use. A human impress applies in reality almost everywhere in the world but the extreme anthropogenic quality of the Netherlands makes it of outstanding interest to the ecological historian.

Marx referred to the original conditions of production that cannot themselves be produced. The Dutch did produce theirs, by more than one act of creation. Despite the stereotypical countryside of polders and windmills, the centuries have seen a series of modifications to the landscape and in wildlife's adaptations. These developments, though patent in outline, are hard to describe in detail because evidence about the initial changes is sketchy. As was said in another environmental context, recovering the ecological history of the Netherlands is 'a bit like trying to reassemble a very complicated mechanism from its parts without having a good blueprint of what the mechanism once looked like.'[iii]

A feature of Dutch success was by-passing the supposed ecological crisis of late medieval Europe by trading heavily on their country's comparative advantage. Trading is the right phrase; it was by commerce that the Dutch enriched themselves. They did not merely transact with their neighbours but positioned themselves at the centre of third-party commerce radiating from Amsterdam as far as North America, Spitsbergen and Java.

Besides illustrating how contingent environmental experience is, the Dutch example addresses five broad issues. First, it raises questions about environmental explanations and their limits; this follows from looking sceptically at conventional arguments that a deteriorating physical environment 'forced' the Dutch to develop an urban, commercial and industrial economy. Nothing 'forced' them to do anything, the environment merely presented a range of costs favouring certain opportunities and foreclosing others. Secondly, it suggests that business history, seldom drawn on explicitly in environmental studies, is an appropriate way of thinking about the drainage schemes by which the Dutch altered their own and other landscapes. Thirdly, it shows that market prices supply but sometimes distort information about wildlife history. Fourthly, while the Dutch case is cited as supporting a declinist view its real history puts this in better perspective. Fifthly, the 'forgotten man' of environmental history — the gain to human society — is highlighted.

The reclamation of peat began after 800 A.D. and dykes to push the sea back, rather than just to stem flooding, were probably first built about 1100. The scale of reclamation gives the lie to the commonplace that the European environment lacks an equivalent of the dramas of expanding into new lands: creating a country from the sea is dramatic enough for anybody. But it happened early and by world standards was small scale, and so is readily passed over. Data on the wildlife effects are sparse and uneven. Dutchmen struggling to 'wrest their land from the sea', which is hardly hyperbole, had little leisure to note these things. They were too busy ditching and draining and harvesting colonial nesting birds, fish, and eels. The clergy, who did have the time, were primarily interested in rental values; their records provide some information about land use but not much about wildlife.

Reclamation proceeded far from steadily, being interrupted by reversals as well as marked by spells of slow-down. The activity of the eleventh, twelfth and thirteenth centuries was followed by a retreat: the sea had already been let in as a result of uncoordinated peat-digging and great dykes would have to be built to drive it out. Whereas damage from the floods of the twelfth and thirteenth centuries had been made good fairly soon, the severe ones of the fourteenth and fifteenth centuries proved almost overwhelming. Agriculture also retreated in inland areas. According to episcopal records, a high proportion of the farms in Twente were 'going back to the wild' in 1385 — the phrase being that of the late Slicher van Bath, doyen of Dutch and probably all European agricultural history.[iv]

The problems were caused by the combined effects of nature and human action. Reclamation in itself could be damaging: the sandy soils uncovered behind the coastal dunes tended to dry out and after about 1400 became subject to erosion. Climatic change, which in the Middle Ages was purely exogenous, is occasionally blamed, but (long before global warming became a shibboleth) Slicher van Bath noted that the climatic explanation could be 'demolished' on grounds of chronology: it was not in depression but in the era of prosperity between 1150 and 1350 that bad weather had been prevalent, with floods, heavy snows, excessive rainfall and (to cap it all) severe droughts.

The Dutch had developed techniques to cope with excess water, as the remarkable medieval dykes and polders show. But when agricultural product prices were low they did not exert themselves to maintain or repair the dykes. Slicher van Bath took no environmentalist view, which had not yet become fashionable during the post-war years. He thought instead in terms of responses to price movements, believing that the reason works of reclamation were neglected and repairs not undertaken during the fourteenth century was simply because they did not pay. The underlying reason seems to have been a drop of 25 percent in the human population.

Authors who put the blame on human action indict the excessive reclamation of peat bogs. This intensified after 1200 and, although there was a pause during the downturn of the fourteenth century, the mining of peat for fuel resumed between 1400 and 1600. It created huge excavations that filled into lakes. After 1530 a technology then new to Holland made it possible to dredge up peat that actually lay under water. This was the baggerbeugel, a long-handled wire-mesh net with an iron lip — Albert Niphuis of the farm Kerkewoning near The Hague showed me the one he still uses every year to clean 400 metres of ditch. The occupier of his farm had introduced the tool for harvesting peat in the early seventeenth century. We will discuss its origins in the next chapter.

Larger lakes became a feature of the seventeenth century. Nienhuis's comprehensive environmental history of the Rhine-Meuse delta tells us that they 'must have' (dangerous but scarcely avoidable phrase) increased the habitat for freshwater fish, especially eels, 'although data are only scarcely available' (sic). Indeed the data amount only to a handful of archaeological findings. After that period, information on land use becomes more plentiful.

In 1600 much of the western Netherlands was new land.[v] Mud flats on the coast had been dammed up and marshes behind the dunes had been empoldered (converted to drained fields). As is often observed, maps of the era show North Holland taking shape amidst a maze of lagoons. Medieval developments had picked up strongly again in the sixteenth century. They faltered and ceased during much of the Eighty Years War, although between 1608 and 1612 (approximately corresponding to the early part of a formal interlude in hostilities), Amsterdam merchants financed the drainage of the Beemster Meer and built themselves summer houses. Nevertheless, half or two-thirds of the way through the seventeenth century reclamation again went into remission and the merchants retreated to the city. Beemster soils had not proved to be as fertile as they had expected. During the first half of the eighteenth century yet another downturn in reclamation was evident, though by then the familiar Golden Age landscape had been established.

Average area poldered per annum, in hectares	
1540–64	1474
1640–64	1163
1740–64	404
1840–64	1568

After Slicher van Bath, *Agrarian History of Western Europe*, p. 200

As early as 1400 it was becoming clear that draining enough land for profitable grain growing was not going to be affordable. Land reclamation, technically impressive though it was, came up against cost limits. The peat shrank and the land surface subsided, while the water table rose. Massive peat-digging for fuel later exacerbated the problem. Between 1600 and 1800 over 100 lakes appeared — flooded peat diggings with a total surface area of 600 square kilometres. Windmills brought some flooded territory back into production, but chiefly as pasture. The important commercial products were wool, butter and cheese, not grain.

Windmills to pump water off the land were the key technological advances and dated from the fifteenth century. A study of the area around Nieuwkoop, east of Leiden, shows that it was windmills which made it possible to polder the land.[vi] Farmhouses were built close together in an orderly row, each backed by a meadow approximately 100 metres wide and as much as one kilometre long. At some point another tranche of strips was started; on the map or from the air this all makes for a rectangular landscape. Cereals were sown for household consumption but most of the fields were grazed by sheep to supply wool for the cloth industry in Leiden.

More powerful windmills were introduced by the seventeenth century. Their primary function was to drain water off the land. The scale of activity may be gauged from the fact that there were eventually 10,000 windmills. 1,200 survive, the biggest concentration today being 228 in the province of South Holland. On the land drained, dairying proved more profitable than growing cereals. Moreover wood for domestic heating was very scarce and the solution was to dig peat, so the turf could be burned. Peat could be mixed with ashes and used as fertiliser on crop land but city demand made selling it for fuel more profitable. It was more valuable or cheaper to produce than grain, alternative supplies of which could be imported through the Sound from the extensive Baltic lands. Peat production therefore continued apace.

The importance of peat as fuel for household and industrial purposes was great. From Nieuwkoop peat was sent by canal to the turf brug, a bridge that still stands near the centre of Leiden. With so many waterways there were few places where it was necessary to use carts. Canal transport reduced the need to allocate land for growing feed for horses, which was a special advantage to the Dutch economy. Astonishingly, peat was shipped to England in the sixteenth century, though almost certainly in tiny quantities.[vii] Perhaps it was a backload in vessels that would otherwise have returned to England empty.

Peat from Holland was also exported for domestic heating to Antwerp, Flanders, where local sources had been worked out. Fuel-scarce Flanders did plant woodlots too, even though at the expense of the acreage available for farming.[viii] Switching land uses was not lightly undertaken in less-developed economies and indicates the enormous importance of fuel supplies. After the late eighteenth century, coal replaced firewood on a sufficient scale for Flemish woodland to be turned back into farmland. There was a later spell of afforestation but in recent times coppicing has been abandoned, with losses among ecotone species. As a result of this fluctuating history, most remaining Flemish woods are mosaics of plantings of different periods, with very small genuinely ancient patches. Land use is plainly dependent on market demand that changes over time.

Around Nieuwkoop and in similar districts peat digging exacerbated the problems of drainage. Between the long strips of the diggings only narrow skeletons of land were left and they were too insubstantial to withstand wave erosion. The result was larger and larger lakes. Van Zanden notes that peat was initially dug close to the cities, according to the Ricardian logic that the most productive resources will be exploited first. This abstracts from land ownership — it does not allow for nearby landowners who might resist exploitation or others further away eager for development — but the logic is reasonable. Externalities were not taken into consideration and large lakes came into existence threateningly close to Amsterdam and Leiden; the vast Haarlemmermeer close to the former happened to be located so that the prevailing winds drove in waves that eroded the shoreline. Despite some fishing, these lakes were not very productive.

Land reclamation has had ups and downs over the centuries. Even with advanced types of windmill to pump out the water, the soaring marginal costs of drainage restricted the crop land that Holland could make for itself. During the second quarter of the seventeenth century clearing sand off arable land in the line of villages all the way from Haarlem to The Hague was able to add some acreage, but not enough. Plans to drain the Haarlemmermeer had to be abandoned because this could not be done until steam pumps became available in the mid-nineteenth century.

Fresh environmental problems added another dimension. De Vries and van der Woude observe that by 1670 the economy had adjusted to the limitations of its resource endowment by turning to trade and industry (as befits economic historians, they say adjusted rather than 'forced'). The new problems of the silting of harbours and industrial pollution of water were unintended consequences of this development. Pollution was a case of industries ignoring the negative externalities again, since textile and beer production depended on copious supplies of the clean water they were busy dirtying. Industrial water pollution continued to be vexatious into modern times.[ix]

THE IMPACTS OF DRAINAGE AND EXPLOITATION OF WILDLIFE

> 'In Holland, where I grew up, my ancestors had killed all the worthwhile game, so I couldn't become a hunter.'
>
> Niko Tinbergen, ethologist and Nobel Laureate

Recent studies of environmental change on the Dutch coast and estuaries have been very thorough. The synthesis by Heike Lotze and collaborators is excellent, though there is greater detail, particularly about birds, in the volume by Piet Nienhuis.[x] Archaeological, geological, historical and biological expertise is called on in both publications. Here I shall continue my practice of considering the impact of changes on birdlife, which exemplify the main ecological effects, though to be fair — and perhaps surprisingly — some organisms lower down the food chain may have fared worse. Beyond reasonable doubt, the destruction of wetland habitats was the main force affecting birds, though losses or reductions can rarely be attributed to this cause in a historically documented way. Most (but still patchy) information is available on the catching or killing of particular species.

During the period from 800 B.C. to A.D. 1050, the low human populations around the Wadden Sea seemingly had limited impact. They wielded weak technologies. Pelicans and flamingos 'perhaps' disappeared, but even if this is true, the archaeological record does not permit us to tell whether human activity was the cause. The Lotze and Nienhuis studies are not immune from the habit of adding conjecture to the evidence. Their suppositions are plausible because they are broadly based but they still involve extrapolation.[xi] Nienhuis, for example, presents a diagram (Fig. 19.1) of the 'assumed composition' of the avifauna of his area over the past 7,000 years. It rests on fusing knowledge about birds found in modern habitats with estimates derived from palaeo-geographic maps of the area of similar habitats in the past. This is logical but very remote from actual observation.

The Lotze study claims that between 1050 and 1500, with increasing human numbers and slowly improving equipment, together with coastal drainage, a number of large bird species decreased. Between 1500 and 1800 still more birds (and wetland habitats) declined. Nienhuis, on the other hand, thinks that as early as the Middle Ages all water birds peaked in number, except herbivores, which increased seven-fold thanks to the continued expansion of dry 'billiard cloth' pastures. The nineteenth- and twentieth centuries (until 1970) were when 'most' bird species were lost, according to the Lotze report. Species that succeeded in remaining on the local list often saw their populations shrink markedly. Nienhuis says, however, that the species present in the delta have not changed much over the past seven thousand years, although their total biomass has fallen. The greatest phase of decline, as with organisms in general, came about 1900.

The authors of the Lotze study and various contributions tributary to it are not blind to counter-movements. They observe that the upsurge of draining and diking in the late Middle Ages created new habitats that favoured certain species — drier grasslands are said to have encouraged the black-tailed godwit and lapwing — which is consistent with Nienhuis's opinion.[xii] Yet lapwings do not like improved grassland and in more recent centuries have declined after reclamation and drainage. They are, and especially were, marshland nesters, although able to switch to cultivated fields at lower densities; there is a study which reports this as the sequence of events in Switzerland after 1880.[xiii]

The Lotze study considers that regulations to control exploitation, which emerged about then, perhaps slowed but ultimately could not prevent the decline of species that were hunted. The assumption is that growth in the human population overwhelmed efforts designed to sustain the 'harvest' of wildfowl. No direct evidence is presented on this crucial point, merely a statement that in medieval and modern centuries intensive exploitation caused herons, cranes, spoonbills, cormorants, duck and geese to decrease. Little egrets were exploited too; large colonies were mentioned in fourteenth-century documents ('the birds were harvested and sold'). They are now rare breeders.

The golden plover exemplifies the variety of human influences on birds. It probably benefited from drainage changes, since it feeds on earthworms, which became more accessible once the grassy polders had been established. In recent centuries the species nested in the eastern and southern Netherlands on heath land, where the vegetation was kept short by sheep grazing and regular burning to improve pasture quality. Golden plover nested only where the heather had been burned from three to eight years earlier, showing how closely it was tied to land use practices. (Similar practices made the uplands of Britain suitable for nesting golden plover; the original upland forest having been unsuitable.[xiv]) When sheep farming and the associated burning of heather dwindled away in the 1930s, the Dutch breeding population was lost.

Golden plover may have been caught on a sufficient scale to form an important item in the diets of medieval Dutch peasants. Wilsternetting (netting waders over decoys; Wilster is Fries for golden plover) is depicted in a painting of North Holland in 1616 and may be far older, as well as widespread in Europe. In the Netherlands legal restrictions had begun to surround the practice during the nineteenth century, but it was still permitted in Friesland to take golden plover for sale.[xv] Interestingly, a law of 1852 tried to ban netting, in order, in a roundabout way, to protect the trade in lapwings' eggs, though lapwings continued surreptitiously to be used as decoys.

In the early twentieth century the catch from wilsternetting is estimated at 100,000 per annum. Before the Second World War most were exported to England. Lapwings, caught illegally in the northern provinces of the Netherlands, were also exported to England under the thin disguise of 'Dutch plovers'. Both species appeared on the menus of expensive restaurants in Brussels, Paris and especially London. During the Second World War the number of catchers supplying golden plover to the Dutch market rose steeply, which is not surprising given the shortage of food, notably in the Hongerwinter of 1944–1945. Catches fell away once prices decreased. During the early post-war years the take may have been in the range of 40,000 to 80,000, declining to 18,000 in 1969. These were migrant birds. The consensus of opinion is that it still meant over-exploitation, since the bird's overall population was falling. Wilsternetting was halted in 1978 and shooting was banned in 1993. Denmark had banned shooting even earlier (in 1982) but killing golden plover for fun is still permitted in England, France, Spain, Morocco and Algeria.

Dutch political opinion has sustained some forms of exploitation in order to respect cultural traditions, which is why wilsternetting lasted so long. A survey of historical exploitation by species is provided by J. T. Lumeij and his associates.[xvi] For instance, falconry was traditional, with a number of prey species. Until feudal rights were abolished in Napoleonic times large species of birds could be hunted with falcons only by the upper classes, after which there was a free-for-all. At Valkensvaard a great annual auction long continued of the raptors that had been taken on passage across the heaths, often using great grey shrikes as decoys. The netting of geese is described from the fifteenth century, wilster netting from the middle ages, and woodcock netting from the sixteenth century. Quail were also netted. The nets for woodcock were set between trees but this was worth doing only when there was a large fall of migrants and is now carried out solely for scientific purposes.

The same change of purpose applies to the few duck decoys still in operation under licence. The decoy was a Dutch invention of the middle ages. Around the Wadden Sea 142 duck decoys have been counted, of which 29 still exist, 13 are derelict, and 100 have disappeared.[xvii] Each decoy caught between 1,000 and 12,000 birds per annum, giving a total of 100,000 to 200,000, though in addition estuaries in the south-west of the Netherlands were active. The chief species taken were mallard, teal, wigeon and pintail. A few licences are still issued for taking mute swans, another ancient practice.

Eggs, except those of swans, could always be collected legally and those of lapwing were once taken in numbers. Friesland is the only place in Europe where this is still permitted — on the grounds of cultural history — during a season lasting from the 1st March to 9th April. A total take of 5,939 eggs is allowed. Permission to take each egg has to be sought by text-messaging and only first clutches may be collected, after which the collectors turn themselves into wardens to protect second clutches. Gull and tern populations were suppressed by fowling and egg-collecting until the flood disaster of 1953 stopped egg-collecting abruptly: 10,000 eggs had previously been harvested per annum. Eijerland (Egg-land), now a polder in the north part of the island of Texel, formerly sent the bakers in Amsterdam 30,000 eggs every year. Other nesting areas were harvested and egg collecting supposedly led to the (local) extirpation of several species on Texel.

English experience confirms the way in which commercial egg-collecting reduced populations. During the nineteenth century lapwings' eggs were sold on a scale that shows they were then a regular food, not just a delicacy. Gulls' eggs filled the same role and clearly the harvesting of their eggs outlasted the peak takes of plovers' eggs. A table of relative palatability drawn up by an expert panel gave the black-headed gull's egg a score of 88 percent against the lapwing at 100 percent.[xviii] Lapwings' eggs were taken extensively in East Anglia: in a single year one egger took almost 2,000 eggs near Potter Heigham on the Norfolk Broads. On an estate near Thetford in the Norfolk Brecks 3,360 eggs were collected in one year during the 1860s, 720 in the 1880s, 72 in 1902, so that by 1925 only twenty pairs were nesting. The clouds of lapwings that bred in the mid-nineteenth century can be only imagined. But the toll was exhausting and lapwings' eggs, or eggs sold as lapwings', were supplemented by imports from Ireland, Denmark, and especially Holland.

The greater avian productivity of Dutch wetlands was evident in the way part of the English market was supplied by wildfowl from Dutch decoys. These eventually invaded the markets of Norwich itself, despite the city's proximity to the bird-rich Norfolk Broads. A Norfolk dealer, who had sent a consignment of 400 duck of various species, 500 snipe and 150 golden plover to London in the winter of 1829, was disappointed to receive only a derisory payment from the London dealers when he despatched large numbers of snipe to them later in the nineteenth century.[xix] The payment was accompanied by a note saying that cargoes of snipe were coming over with wildfowl from Holland.

As to passerines, lark mirrors were used in the Netherlands to attract skylarks to be netted. Estates in the coastal region possessed finching yards (vinkbaanen), where finches were netted on a large scale as early as the middle ages. During the seventeenth and eighteenth centuries this became a pastime of landowners, who took chaffinches, siskins, goldfinches, twite and bramblings, which were all thought to be delicacies. The practice was at its height in the late eighteenth century, with many thousands of birds taken every year.[xx] The harvesting of this great migration perhaps did less harm than might be expected, merely removing a surplus that would not have survived to breed the next season, leaving the prime sites and food supplies for the survivors. Chaffinches still pour down the dune coast from Scandinavia each autumn, en route to Spain or (seemingly the females) to England.

Wolff, in his study of exploitation around the Wadden Sea, calculates that five species of birds have become extinct during the past 2,000 years, and that as with other organisms the main causes were habitat destruction and over-exploitation.[xxi] He believes that the heaviest pressure on natural resources probably came during the nineteenth century. After about 1970 several species of waterbirds began to recover. Eider duck, lesser black-backed gulls and common gulls, all of which had apparently been driven out by egging, returned to Texel as breeding species in the twentieth century. Naturalists may be over-excited by this since their baseline tends to be only two or three generations ago, when bird populations were particularly low and against which a limited recovery is certain to be conspicuous. Nevertheless conservation is having a positive effect and the continued high growth rates of some species (the White Stork is one) may mean that they have not yet reached the current carrying capacity of the Dutch environment.

In sum, the message with respect to water birds is not wholly discouraging. Leaving aside differences of opinion among authorities about the timing of peak exploitation — differences that are not easy to resolve given the available data — it seems that the most significant depressants were over-exploitation and habitat change. Apportioning the effect between these forces seems an intractable problem. Exploitation has been substantially reversed, that is to say forbidden or at least curtailed. Habitat change is reversible in principle but would be astronomically expensive on a large scale and touches on too many interests to be contemplated wholesale. Lotze declares in another article on the Wadden Sea that the evidence shows 'negative trends can be turned around if protection and restoration are integrated into ecosystem-based management plans.'[xxii] An economist might be forgiven the aside that, despite appearances, the losses of past centuries were not bought at an abnormally high price, given the subsequent recovery of some species and the social gains secured by past exploitation.

Even in the Netherlands, coastal and marshland birds are only a proportion of all species and whether similar optimism is justified about the entire avifauna is debateable. Wild habitats greatly diminished during the nineteenth- and twentieth centuries. Seventy percent of the Netherlands are now farmland, with the highest yields of all European Union countries, produced by an industrial agriculture unfriendly to many birds. Heathland, estimated at over 600,000 hectares in 1833, was reduced by the mid-1980s to 42,000 hectares and these were 'threatened' by eutrophication and recreational uses.[xxiii]

The term 'threat' springs unbidden to the lips of naturalists but it is a value judgement. The countervailing expansion of suburbs and of the 'green' towns in western Holland has been accompanied by the establishment of hundreds of thousands of nest boxes and feeding stations, with positive effects on a burgeoning range of species readily enjoyable from the windows of ordinary Dutch homes. Clearly there are trade-offs between different sets of species, mainly the unintended results of altered habitats, as well as trade-offs between all other organisms and human welfare. Despite costs in biodiversity, only misanthropes who dislike their grandchildren or at any rate other people's grandchildren, may be rash enough to shout from the rooftops that the tripling of the Dutch population during the twentieth century should be regarded as having done unmitigated harm.

ENDNOTES

[i] I am indebted to conversations with Petra van Dam, Tijs Goldschmidt, Robert de Groot, the late Frans Husken, Vrouwke van Marion, Albert J. Niphuis, C. A. Overdevest, Barto Piersma, Peter van Rooden, the late Bernard Slicher van Bath, Jo Spaans, Jan de Vries, and the late Ad van der Woude.

[ii] For a map of this area, see Bert de Vries and Johan Goudsblom (eds.), *Mappae Mundi: Humans and their Habitats in a Long-Term Socio-Ecological Perspective* (Amsterdam: Amsterdam University Press, 2002), Fig. 11.5, p. 387.

[iii] Callum Roberts, quoted in the *New Scientist*, 28 April 2012.

[iv] B. H. Slicher van Bath, *The Agrarian History of Western Europe A.D. 500–1850* (London: Edward Arnold, 1963), p. 160.

[v] For the history of drainage in the Netherlands, see for example Violet Barbour, *Capitalism in Amsterdam in the 17th Century* (Ann Arbor, Mich.: Michigan University Press, 1963); Piet H. Nienhuis, *Environmental History of the Rhine-Meuse Delta* (Springer, 2008); Slicher van Bath, *Agrarian History*; Petra van Dam, 'Sinking peat bogs: Environmental change in Holland, 1350–1550,' Internet version, n.d.; de Vries and Gousdblom (eds.), *Mappae Mundi*; and Jan Luiten van Zanden, 'The Ecological Constraints of an Early Modern Economy: the Case of the Holland (sic) 1350–1800,' in S. Cavachiocci (ed.), *Economia ed Energia secoli XIII-XVIII* (Florence: Le Monnier, 2002), pp. 1011–1030.

[vi] Peter Westbroek, in de Vries and Goudsblom, *Mappae Mundi*, Chapter 11.

[vii] C. T. Smith, 'Dutch peat digging and the origin of the Norfolk Broads,' *Geographical Journal*, 13/1 (1966), p. 69.

[viii] K. Kirby and C. Watkins (eds.), *The Ecological History of European Forests* (Wallingford: CAB International, 1998), p. 283.

[ix] K. J. W. Oosthoek, 'The stench of prosperity: Industrial water pollution in the Groningen Veenkolonien,' Environmental History Resources website, University of Newcastle, n.d. [2011]

[x] Heike Lotze *et al.*, 'Human transformations of the Wadden Sea ecosystem through time: a synthesis,' *Helgoland Marine Research*, **59** (2005), pp. 84–95; Nienhuis, *Environmental History*, pp. 510–535.

[xi] Compare the similar synthetic method of Louwe L. P. Kooijmans, 'Wetland Exploitation and Upland Relations of Prehistoric Communities in the Netherlands,' Internet version, p. 115, and his conclusion that the range of bird species was not drastically unlike the modern one, though populations were much larger.

[xii] A continued element of surmise among writers is revealed by Andrew Wareham when discussing the East Anglian fens, on p. 18 of Hilde Greefs and M. C. Hart, 'Water Management, Communities and Environment. The Low Countries in Comparative Perspective, c.1000–c.1800,' *Jaarboek voor Ecologische Geschiedenis 2005/2006* (Gent: Academia Press, 2006) [Fully available on Google Books]. Wareham claims that swans, duck and heron were introduced (sic) in place of curlew and black-tailed godwit after the medieval embanking of wetlands. This seems neither ecologically likely nor consistent with Dutch opinion.

[xiii] Richard Vaughan, *Plovers* (Lavenham, Suffolk, 1980), pp. 37–38.

[xiv] Vaughan, *Plovers*, pp. 37–38.

[xv] Joop Jukema *et al., Goudplevieren en Wilsterflappers* (Utrecht: KNNV Uitgeverij, 2001). English summary, pp. 237–240.

[xvi] J. T. Lumeij *et al., Beter een vogel in de hand. . .* (Zeist: KNNV Uitgeverij, 2008), pp. 187–189.

[xvii] Wim J Wolff, 'The exploitation of living resources in the Dutch Wadden Sea: a historical overview,' *Helgoland Marine Research*, **59** (2005), pp. 31–38.

[xviii] Vaughan, *Plovers*, pp. 126–127.

[xix] Arthur H. Patterson, *Nature in Eastern Norfolk* (London: Methuen, 1905), p. 72.

[xx] This is the subject of a massive quantitative study by Ignaz Mathey, *Vincken Moeten Vincken Locken: Vijf Eeuwen Vangst van Zangvogels en Kwarttels in Holland* (Hilversum: UItgeverij Verloven, 2002).

[xxi] Wim J. Wolff, 'Causes of Extirpations in the Wadden Sea, an Estuarine Area in The Netherlands,' *Conservation Biology*, **14** (3) (2000), pp. 876–885.

[xxii] Heike K. Lotze, 'Rise and fall of fishing and marine resource use in the Wadden Sea, southern North Sea,' *Fisheries Research*, **7** (2007), p. 217.

[xxiii] Rob G. Bijlsma *et al., Algemene en schaarse vogels van Nederland* (Haarlem: GMB Uitgeverij, 2001), p. 8.

9. ENGLAND: RECLAMATION AND EXPLOITATION

> There is something much more obviously man-made about Lincolnshire than anywhere else in the country; you know that the Romans dug some of the drains, then the Dutch. Vermuyden's drain anyone?
>
> John Aston (2008)

Dutch drainage engineers altered wetlands in a remarkable number of places right across Europe. Their projects are mapped as a double-page spread in Charles Wilson, *The Dutch Republic*.[i] Before 1250, land reclaimers from the Leiden area were at work in North Germany, having learned their trade by digging peat, since even at home actual dykes were not being systematically built so early.[ii] They could, however, offer technical skills and know-how about organising the water management of village communities living free from manorial control.

Not every wetland to which the Dutch travelled was successfully drained at the first attempt but most succumbed eventually and it is safe to assume that much natural abundance vanished. Concentrating on the major reclamation projects may actually distract attention from the total scale of drainage that has taken place over the last few centuries, for there was an enormous number of lesser efforts and in the nineteenth and twentieth centuries a widespread under-drainage of fields using industrially-made clay 'tiles' or pipes. Europe's ecology was dried out. The continent had once been dotted with marshes full of wildlife of which modern animal, bird, fish and plant populations represent only fragments.

England was an early and extensive employer of Dutch drainage services. References to Cornelius Vermuyden's scheme to drain the Great Level of the Fens are *de rigueur* in the literature of the topic, although it is harder to learn about the work of other Dutchmen. Moreover, allusions and longer treatments admire technical feats of civil engineering almost to the exclusion of the context. They imply that the Dutch began with a monopoly of the relevant skills: 'drainage was taught them by the Dutch' is a typical phrase, though it exposes the fact that know-how did not remain a permanent monopoly even if it really was one at the outset. The diffusion of technology has always been hard to frustrate.

The Thames estuary and the Zuider Zee (IJssel Meer) were physically alike, and during the Middle Ages contact between the inhabitants of their shores was frequent, as was contact with ports in northern France.[iii] Eel boats from Holland were visiting the Thames in the thirteenth century, and fishing methods usually attributed to the Dutch were adopted or perhaps even invented there. Notable among them was what the English called the 'Wondrychoun', or drag net, possibly used by the fishermen of Barking as early as 1405 and probably the same as the Wonderkuil of Holland. It was still being used in the Netherlands after the First World War to scoop up small fish which were fed to 170,000 duck and 25,000 hens annually, indicating Dutch specialisation in intensive poultry farming.

Dutch enterprises did have comparative advantage in drainage work and Dutchmen were the people most likely to be hired to undertake it. Their advantage rested on much besides engineering, including small 'p' political skills in organising drainage settlements and the liquidity of Amsterdam's capital market. Engineering leaves visible and tangible remains, whereas the organisational skills of past times have left traces only recoverable, where recovery is possible, from old books and business archives. The subject of drainage is properly an episode in business history in which Dutch entrepreneurship and capital played a central role. Without supporting attributes, mere engineering could never have extended the Dutch reach across early modern Europe.

Another reason for approaching the subject from a business history viewpoint is that England, at any rate, did not strictly need to import techniques of drainage and flood prevention. England was at a relative rather than absolute disadvantage. Techniques of draining and reclaiming were already known. Roman drains and seawalls may have become virtual fossils by medieval times, but there were subsequent native efforts at draining land and erecting flood control works. In the Middle Ages, the church undertook the bulk, at least of the larger ones. Cistercian drainage at Byland Abbey, Yorkshire, was said to have constituted the largest English water engineering project at the period. Other examples include Old Splott Rhine, which passes into the Severn estuary through the Vale of Berkeley, Gloucestershire. This was a ditch dug in 1346 by engineers who had not only to contend with the terrain but find a way to settle a dispute between the village of Elburton and the abbot of Malmesbury. In the Fens, a major drainage cut was organised by Bishop Morton, of Morton's Fork, who became Henry VII's Chancellor of the Exchequer.

The clyse at Highbridge, Somerset, dates from 1485, although the structure has been replaced by one erected in 1802: it is a sluice to keep out high tides, preventing the river Brue from backing up and flooding the Somerset Levels. Clyse is a local term which somewhat unexpectedly shares its derivation with sluice, both words ultimately meaning closure. Of the rivers through the Somerset Levels, only the Parrett lacks a clyse. Furthermore, to reinforce an appreciation of native enterprise in hydraulic engineering as well as the international market in associated skills, consider the one-and-three-quarter miles of that very early canal, the Exeter Ship Canal. Devon merchants employed John Trew of Glamorgan to undertake its construction and in the mid-1560s he installed the first pound locks in the whole country. Later, in 1671, the Exeter authorities consulted a Dutch engineer about dredging the canal although in the event they chose Richard Hurd of Cardiff. There was a long list of English engineers responsible for water projects right from 1500 — a list is supplied in the *Biographical Dictionary of Civil Engineers.*[iv]

Some accounts of English economic history play down native involvement in favour of portraying an underdeveloped country whose chance of economic progress depended on imported foreign enterprise. There were of course incomers and during the Golden Age of Holland they were often Dutchmen. English backwardness is exaggerated by concentrating on the role of immigrants, right from the invitation to the Lombards to settle in medieval London and bring their financial practices with them. In agriculture, the Dutch contribution was noteworthy but even there can be over-emphasised. One specialist study becomes guarded when discussing influences on pastoral farming, the evidence for direct Dutch transfers of livestock and livestock management practices being circumstantial.[v]

Too little allowance is made for independent discoveries in similar environments. This is easiest to show by citing an example a little outside agriculture. The baggerbeugel, mentioned in the previous chapter, was used in Holland soon after 1500. But the identical tool had been used in Norfolk from the late thirteenth or early fourteenth centuries, under the name of 'didle'. A student of the topic implies that the flooding of peat beds 'forced' Norfolk men to invent this tool and that when the peatlands of Holland became liable to flood, rather later in time, its use became obligatory there too.[vi] This is another instance of the misconception we found in the Dutch literature, that a need 'forces' the appearance of whatever solution historians happen to observe: necessity is the mother of invention, except of course when it is not, for hats do not always produce rabbits! The same author concludes, rather helplessly, that the didle or baggerbeutel was either introduced to Holland from England or was evolved independently by the Dutch. The sources appear to admit no determination.

Hence techniques imported to England may not have been as formative as convention makes them seem.[vii] There were schemes of wetland reclamation in which the Dutch were not involved and drainage schemes long before Vermuyden was brought in.[viii] The Dutch role is hard to distinguish from native English enterprise because Dutchmen had colonised the Fens around the Wash back in the twelfth and thirteenth centuries. Vermuyden's own seventeenth-century scheme for the Fens followed plans drawn up by John Hunt a generation earlier, in 1604–1605, and there were other English civil engineers in the trade.

Dutch achievements must certainly be recognised, while noting that what the Dutch themselves borrowed is rarely discussed. For instance, the fodder crops cultivated so intensively around late medieval Dutch towns came from Italy and sometimes before that from the Middle East or even India. As far as England's genuine debts to the Low Countries are concerned, a Hollander was paid for work on the sluice at Romney as early as 1410; Flemish masons were employed in the fifteenth century on the sluice and dam at Boston, Lincolnshire; and an Act of 1536 sanctioned reclamation at Wapping to be carried out by a man with the tell-tale name of Vanderdelf. He was to receive half of the acreage reclaimed. Despite the signs of native ingenuity and endeavour on the part of the English (and Welsh), Dutch influence was evident all along the East and South coasts from the late sixteenth century. This was apparent in architecture, agriculture and especially sea defences. Numbers of Dutchmen immigrated, including entire workforces to drain the Fens. The list of English wetlands tackled with their assistance is long and widely scattered. Vermuyden even bought land on the western side of England, at Malvern Chase, Worcestershire, a section of a common that Charles I had acquired through relinquishing royal rights over the whole area.[ix]

An important advantage possessed by the Dutch was the windmill. Nevertheless the windmill was not new to England, probably having been introduced in the twelfth century by crusaders returning from the Holy Land. The main adopter phase was, however, late by Dutch standards, though by the early eighteenth century, windmills were numerous in the Fens. (Even with this technology not all problems could be solved: the flood gates meant to safeguard the draining of Martin Mere, Lancashire, were destroyed by storms and the project could not be completed until the introduction of steam in the 1780s.)

With so much to-ing and fro-ing, trying to allocate praise between nations is not fruitful. Incomplete sources add to the uncertainty. How, then, should the English record be evaluated? Lawrence Harris has argued that drainage in England was delayed for 40 years because schemes were thrown open to 'unrestricted financial competition'.[x] He thinks that, without the lobbying frenzy this unleashed, Fen drainage might have started as early as 1589. Conceivably jockeying for permissions and capital was tedious but a 40-year lag is implausible: competition was more likely to get things moving than to slow them down. And if Englishmen themselves were shy of investing, there was abundant capital in the commercial capital of the Western world, just across the North Sea, in Amsterdam.

Understandably, the Civil Wars did hold up progress, following a significant burst of activity in about 1630. Risky projects with long gestation periods, requiring the negotiation of multiple property rights, were not suited to wartime. But the phase of country house building that started in 1650 suggests confidence returned immediately after the execution of Charles I, and it is not clear why drainage schemes should not have resumed vigorously too. At any rate, after the Restoration, Dutch models were continually put forward by progressive writers in England, some of whom had experience of the ways of the Hollanders gained by their own wartime exile in the Low Countries. Their countrymen, they felt, needed to be goaded by this example, which was palatable because the Dutch were Protestants and after 1688 a Dutch king came to the English throne.

Accepting Dutch investment involved regular business decisions, at least by the personalised standards of Charles I's reign. The Crown itself sometimes took shares in projects. Standard references to the practice may nevertheless exaggerate the role of individual Dutchmen or imply they were stand-alone promoters. In reality, the capital required for arduous tasks dependent on pick-and-shovel work was so large that collaborators were needed to share the risk. Drainage projects were always being deferred, even reversed, as a result of disputes between claimants to titles over land and water. Sir William Dugdale was paid well for employing his antiquarian skills to strengthen the uncertain legal position of the Duke of Bedford's Company of Adventurers in the Grèat Levels of the Fens. The outcome was his *History of Imbanking and Drayning* (1662). The objectors included communities whose age-old occupations were threatened by the transformation of wetlands where they had been accustomed to hunt, fish, dig peat, or gather reeds and withies. Nothing about this resistance is surprising; it was an anticipatory variant of eighteenth-century dissent over enclosure.

Humphrey Bradley, Brabacon despite his name, had already acquired some English civil engineering experience when he was touting his drainage services there at the end of the 1580s.[xi] He now promised to persuade 'certain gentlemen of wealth' to put up enough capital to drain the Fen, but found no backers. He moved on to France, where he secured from Henry IV a patent for a *Societe pour le Dessechement des Marais et Lacs de France.* Under this grand aegis, drainage was carried out in several provinces, though not always successfully. The key point is that, by the time of his French projects, Bradley had learned to raise adequate finance: each enterprise was funded by a separate group of entrepreneurs, including men from both southern and northern parts of the Low Countries. Chief among them and primarily responsible for keeping the enterprises, as it were, afloat was Jan Hoeufft, who was influential in moneyed and official circles in Paris and had connections in Amsterdam, so that he could draw capital from both centres. During the seventeenth and eighteenth centuries, groups of entrepreneurs obtained concessions for projects in several countries, receiving grants of land in return. They provided advice on hydrology and the building of dykes and sluices but above all supplied capital and negotiating skills. They were employed by foreign monarchs and Van de Ven labels the companies involved the first consulting firms.[xii] Rather than settle on the distant marshes of Europe, as their forerunners had done, technical experts now tended to return to the Netherlands after executing their projects.

In England, investors from Amsterdam and other towns in Holland were involved in several schemes carried out after Bradley's day. One was said to have put £13,000 on his own account into the draining of the Isle of Axholme. Another was a principal undertaker in the Earl of Bedford's company for reclaiming the Great Fen; he was Sir Philiberto Vernatti, originally from Delft despite being both a Scottish baronet and a nephew of the corrupt monopolist of the English glass industry, Sir Robert Mansell.[xiii] In about 1630, Vernatti petitioned for the denization of 14 of his associates, probably all Amsterdammers, whom he had induced to invest in the Fens project. Denization would have made them English subjects, able to buy land, though without full citizenship rights. Their desire for this status shows they feared that away from the United Provinces their legal position might prove insecure. Vermuyden, who was personally related to the investor, van Croppenberg, petitioned in 1630 for the same privilege for 18 men (mostly Amsterdammers) associated with him in draining Hatfield Chase. And East Anglia was in some wise a colony of the Netherlands — the Fens were virtually a drainage colony.

At times, the Dutch associated themselves with local entrepreneurs, for example, one of Vermuyden's relatives was already acquainted with Englishmen keen to have Vermuyden work on Canvey Island, but at other times they parachuted in, so to speak, touting for trade. Opportunities there were because the old monastic responsibility for keeping fen drains open had fallen into abeyance since the Dissolution. Demand therefore drew in Dutchmen. Whether English projectors solicited their aid or joined forces with them, they were hiring the best civil engineering enterprises of the day. To call these enterprises 'firms' might seem anachronistic, since business structures were then more piecemeal and extempore than the term suggests. But the Dutch were the ones best placed to raise capital, and to do so for ventures outside their own country, despite risks greater than at home. The power of the capital the Dutch had accumulated through commerce is shown in other spheres, too, such as the fur trade: the Amsterdam house of Bontemantel advanced so much money far away in Russia that in 1640 almost no furs came on the open market.[xiv] And the Dutch established a real tradition of involvement in overseas wetland and water products, for instance forming a company with Billingsgate merchants as late as 1959 (sic) to acquire exclusive rights to the eel fishery on Lough Neagh in Northern Ireland.[xv]

The Dutch drained and settled in England, France, Italy, Germany, Poland, Sweden and along the southern shore of the Baltic. These ventures took place in the sixteenth century, were most evident during the next century and continued very much later. The earliest ones in the eastern parts of Germany were sometimes matters of religious communities moving to escape persecution. Most of the later ones were commercial exploits, though sometimes responding to invitations from foreign rulers.

EFFECTS ON ENGLISH WILDLIFE

Evidence about the consequences of reclamation for wildlife in England is all too often thin; specific references to birds tend to deal with killing or capturing them for the table. Consider Martin Mere in Lancashire, originally drained in 1697. The author of a study of its landscape and economy relies heavily on depositions made for a court case in 1714.[xvi] The bird most often mentioned is the mute swan, whose cygnets were routinely captured. Geese, too, are known to have been snared but the article tails off into analogies with the Fens of East Anglia. The most explicit sign that wildfowl continued to be exploited at Martin Mere is that a duck decoy was working in the early eighteenth century.

Culinary references tell us most of what little we know about marshland birdlife in the middle ages. Bitterns figured in most feasts. They were always expensive but available at a price. The mute swan population in the Fens has been estimated for the sixteenth century, when the species is said to have been saved from over-exploitation or even extinction by being replaced as a table bird by the introduced turkey. Spoonbills were still on the menu and were less fortunate; the destruction of habitat is held responsible. Breeding colonies in Norfolk are mentioned in 1300 and again in the early seventeenth century but Mark Cocker tells us in *Birds Britannica* that, 'the drainage of the Fens under the direction of the Dutch engineer Cornelius Vermuyden in the mid-seventeenth century *almost certainly* eliminated an important breeding area. By the time Vermuyden's works were completed in 1651, the spoonbill was close to extinction.' (p. 60, italics added) Apportioning responsibility between habitat destruction and the killing of birds rests on a great deal of surmise. Drainage tends to get most blame even though more actual evidence exists about shooting and trapping.

In East Anglia, very large catches of duck were made in decoys used to supply the London meat market. The same was true of Lincolnshire, 31,000 duck being sent from the 10 decoys in the East Fen in one season in about 1800.[xvii] My conjecture is that many of the duck trapped in eastern England no longer nested locally but were winter visitors, once drainage had eliminated many nesting sites. Some will have come from Holland, whence, unluckily for them, the technology of the decoy had been transferred in the seventeenth century.[xviii] Even in East Anglia, the region of England most affected, historical material on decoy catches is scarce but fortunately there is other detailed ornithological information. In the mid-nineteenth century, Richard Lubbock investigated the history of birdlife in Norfolk. His was a position of strength: personal involvement, familiarity with the ground, acquaintance with local gunners, knowledge of specimens taken by collectors, and first-hand investigation of historical sources.[xix] Lubbock, and a generation later his editor, Thomas Southwell, set a standard by documenting their material. Much of what they reported was to reappear in later accounts.

The pertinacity of these authors enabled them to evaluate the developments they witnessed in mid-Victorian times: steam pumps, rail connections with London, more commercial egg-collecting and (they emphasised) better guns in the hands of local fowlers. They considered the history as far back as the Middle Ages. Their texts are stuffed as tightly with detail as any taxidermist ever stuffed a cabinet of birds. 'When first I remember our fens [from 1816]', wrote Lubbock in 1847, 'they were full of Terns, Ruffs, and Redlegs [Redshank], and yet the old fen-men declared there was not a tenth part of what they remembered when boys' (p. iii). On this showing, the marshes had swarmed with bitterns, grebes, ruffs and avocets, to which reports from the Lincolnshire fens also testify. Earlier still, they had swarmed with an even wider range of water-birds, though allowances should be made for the rosy glow of youthful memory.

Southwell noted that, between 1830 and 1880, 20 species vanished or were fast disappearing. The London market was insatiable, yet the number of decoys was likely to fall as the result of continental competition: 'London is inundated with Dutch fowl, which are sold at a very cheap rate; nay, carts may occasionally be seen in the streets of Norwich, loaded with the same commodity from France and Holland' (pp. 146–147). Drainage may have been reducing waterfowl populations in Holland, too, but they were still sizeable enough to supply England. The trade from Holland in species other than waterfowl was also active. In the 1830s, when bearded tits were decreasing in England, large numbers were brought from the Dutch reed-beds every autumn and sold cheaply in the London markets, while in Edwardian times, cages full of Dutch little owls were often to be seen at the Leadenhall market.[xx]

An additional factor depressing the price of duck was competition in the butchers' shops from game-birds slaughtered on landed estates. This was greatly on the increase during the third quarter of the nineteenth century, the notorious 'Age of the Pheasant'.[xxi] Birdlife in historical times was not something that was always naturally present, something that happened as a result of the pure forces of nature; human intervention, both deliberate and unintended, was constantly altering it, never more so than in Victorian times. The meat markets ruled, dictating what it was worth slaughtering and therefore creating a large fraction of the available records. Pressure on waterfowl did not, however, derive solely from this, but from this coupled with the effects of drainage and reclamation. The over-representation of rarities, which tends to distort county avifaunas, reflects a third factor, the obsession of the era with the amassing of specimens for collectors' cabinets.

Recoveries of wetland birds from their historical losses have taken a long time and have been influenced by bird protection laws, changed public attitudes to collecting, and the establishment of reserves. These and other alterations (reversions) of habitat could be afforded only by rich modern societies. Consider the on-off experience of the Boston Wash Banks in Lincolnshire, where reclamation has recently been abandoned. The buildings at Freiston Shore formed a popular seaside resort until 1842, but were left high and dry when a bank was built around a large area of saltmarsh. Banks enclosing saltmarsh even further out were added by Borstal Boys between 1935 and 1979. Their grudging labours pushed the line of the bank so far out into the Wash that no fringe of saltmarsh was left and the bank had to face storm and flood head on. It was breached several times. Today, following a striking reversal of attitudes, it has been deliberately pierced, allowing a protective fringe of saltmarsh to arise on the seaward side.

Recently, there have been heartening gains of marshland birds, involving colonisation or re-colonisation from Mediterranean and Dutch populations of species that, in England, had become extinct as nesting birds. Once before, during the depression of the 1930s, naturalists had hoped that the long history of reclamation might reverse itself, because arable fields were then going back to marsh. Imported grain was so cheap that abandoning ploughland and ceasing to keep the ditches open was a sensible re-allocation of resources, although farmers would not have welcomed this interpretation. A poignant account of one such locality from the naturalist's viewpoint was penned by E. A. Ennion in *Adventurers Fen.*[xxii] His hopes for the recovery of birdlife were dashed during the Second World War. The farmer, Alan Bloom (note his originally Dutch name), reclaimed Adventurers Fen as a contribution to the national programme of land drainage and food production.[xxiii] For decades after the war, when there was no longer a case for producing so much food (imports would have been cheaper), farm subsidies meant that arable use continued. In one place, the recovery of water-birds had nevertheless already begun: avocets nested on an island off the Suffolk coast at the end of the war, when access to the shoreline was forbidden and the birds were undisturbed. Afterwards, deliberate conservation saved the day for them.

Unfortunately, agricultural intensification continued. It is said, for example, that only 3% as much riverside hay meadow survives as was present in 1939. The meadows have been drained and converted into featureless expanses of ryegrass; the buzz of innumerable insects and the call of curlews from the scraps of remaining hay meadow reveal what was lost to the agricultural improver. I remember Mike Soper, the manager of the Oxford University Farm, looking across the Thames in the 1960s and almost literally shaking with an urge to give the ancient Yarnton lot mead a dose of fertiliser. He was a famous agriculturist and people like him usually won the battle for intensification, although not at Yarnton.

The birds that lost out to the drainage of meadows in the nineteenth century, and the still more vigorous reclamation of the late twentieth century, were species like the lapwing and snipe, once so numerous. They also suffered from commercial egg collecting. Snipe numbers rose and fell with the fluctuations in farming prosperity, the falls following losses of the wet patches of grassland that they require.[xxiv] A sprinkling of nature reserves cannot fully compensate for the massive loss of wet meadows.

The history of waders and waterfowl during the nineteenth century, already mentioned with respect to the Netherlands, is nowhere easy to document in a consistent fashion, but this is a convenient point at which to introduce further English testimony and attempt a summary. Typical modern writings refer to a generalised past or bestrew their accounts with a handful of seemingly hard numbers from printed sources. In truth, it is difficult to do much more. With descriptions of the situation in the Norfolk Broads and on the adjacent coast in mind, I would guess that the peak of exploitation was reached in the early 1870s. After that, two or three factors coincided to bring some, though not much, relief — cold comfort since wildfowlers' lamented their prey had largely melted away. Scepticism about hindsight is often justified; here, though, it may be suspended. Even on the remaining undrained land there has been a reduction of conspicuous species.

The ranks of full-time wildfowlers fell through employment opportunities arising elsewhere, higher earnings, or both. Allusions to fowling as a game for old men, who had never known anything else, become more frequent in the literature. One hint that living standards were improving by the end of the 1870s was the abandonment of the Norfolk habit of feasting annually on mistle thrush pies.[xxv] Bird protection began on some private properties, helped by the not wholly ineffective laws of the 1870s.

More explicit information exists with respect to certain sought-after species, lapwing, black-headed gull and a few others. Lapwings' eggs were the most cherished and were said to be worth more than the bird. The species is not a colonial nester but there were places where nests were so dense it may as well have been. At one time nests were thick on the ground in marshes and wet meadows — surely more densely than the maximum of six pairs per hectare cited in the ornithological literature. Had this not been so, the hundreds of eggs one man could find in a morning would have been an impossible take. These high-yielding marshland densities may of course have occurred where nesting areas had been squeezed by drainage and reclamation.

Even on the dry chalk uplands, lapwing numbers could be immense by modern standards. A. C. Smith claimed there was no regular trade in Wiltshire but nevertheless local hotels were supplied and, despite his own hesitation, he noted 40 nests destroyed by farm implements within one week on a single patch of rough ground on the edge of Salisbury Plain.[xxvi] In neighbouring Hampshire, Dewar described the late nineteenth-century egg-gatherer, 'walking up and down the favourite laying-places in April, with his eyes fastened on the ground', seizing the eggs and marking any incomplete clutch, 'to return in due course and empty the nest'.[xxvii] Dewar had found 20 lapwings' nests emptied in a field of 2 or 3 acres. As is often the case with anecdotes like this, their representativeness is unknown. Dewar may have chanced across a greedy individual with no thought for the morrow.

Patterson lists lapwings among the many species on a game-stall at Great Yarmouth in the 1890s.[xxviii] Although their numbers were right down, they remained the commonest 'game' on offer, next to wood pigeons. The lapwing's decrease is plain. It may be deduced from the fact that by the 1870s, coots' eggs, once ignored, were being taken as a substitute, as many as 500 or 600 in a season from one smallish pool near Norwich.[xxix] The statement was made that plovers' eggs no longer repaid the searching, whereas coots' nests remained plentiful and were often conspicuous. Lapwings were still netted and shot in the Fens in the 1930s. After the war, rather feeble efforts were made at protecting the bird. The House of Lords' debates called on Dutch evidence as to whether egg collecting reduces lapwing numbers, but the testimony was inconclusive.[xxx] The House of Lords, doubtless full of men fond of eating the eggs, passed an amendment to the Protection of Birds Act, 1954, permitting the taking of lapwings' eggs before 15 April each year.[xxxi]

In the mid-twentieth century, repeat nesting was often allowed for; tractor-drivers marked nests with a stick and took care to avoid them, though many were still rolled. Today the eggs can hardly be spotted from perches in the vast cabs of modern tractors, even were busy contract ploughmen inclined to look out for them. Licences are still issued to permit the taking of lapwings' eggs for food provided this is done before 15 April.

I have not been able to learn how many such licences are issued, despite enquiry. In the case of another affected species, the black-headed gull, it took a Freedom of Information request for anyone to elicit the information.

The eggs of marsh breeding waders other than lapwing were also extensively taken for sale during the first half or three-quarters of the nineteenth century. Redshank and snipe were the main targets and their eggs were substituted for lapwings', though judged inferior. These species had the disadvantage that they were less likely to lay again, meaning that exploitation hit their populations harder. Redshank, once numerous on the Broads, had deserted their former breeding grounds by the 1870s.[xxxii] The redshank recovered late in the century by spreading inland to the Midlands and there are figures to mark its increase as a result of the expansion of rough pasture during the 'long depression' until 1940. Afterwards, pasture improvement brought about another decrease.[xxxiii]

A problem with records of egg collecting is that the prime alternative to the lapwing, the black-headed gull, went under several names historically, one of which was 'pewit gull'! Colonies definitely of black-headed gulls were rented out in seventeenth-century Hampshire and Staffordshire, while eggs from Norfolk were sold in London.[xxxiv] Figures are available for eggs and sometimes for young birds that were taken. Proprietors of the colonies regulated the eggs to be removed from each clutch and limited the harvesting season, though the rules were not always observed. An average year at Scoulton, Norfolk, was producing over 30,000 eggs in the 1880s and a few years earlier had yielded 40,000.[xxxv] The exploitation was far from casual. Breeding colonies had long been valuable properties and their management was by organisations that were, to all intents and purposes, firms. This was true as early as the reign of Charles I, when 'Pewit Island' in Portsmouth harbour earned its owner £ 40 per annum from the sale of 'pewit gulls', i.e., black-headed gulls, which may have referred to either or both young birds or eggs.

The black-headed gull is the chief species whose eggs have sustained exploitation to the present day. Despite (according to C. A. Johns) not being acceptable to the most fastidious Victorian palates, gulls' eggs were often substituted for lapwings'.[xxxvi] Most 'plovers'' eggs sold in London were said by ornithologists to be in reality those of other birds, usually the black-headed gull, which is a true colonial nester whose eggs are fairly easy to gather.[xxxvii] About 1940, the Leadenhall market is said to have been handling an incredible 300,000 eggs per annum.[xxxviii] In the early twenty-first century, 40,000 gulls' eggs per annum are still consumed by fashionable Londoners, maybe people who have never eaten plovers' eggs.[xxxix]

Despite the evidence of long-run habitat and population decreases of wetland species, there have been some remarkable gains within the past couple of decades, connected with the establishment of reserves. A spectacular array of herons has returned to southern England.[xl] Spoonbills, bitterns, purple herons and other species have come back to East Anglia. Cranes nested in Suffolk in 2007 after several centuries' absence. Purple herons nested in England for the first time at Dungeness in 2010. Flocks of glossy ibis are now regular on southern English wetlands; little egrets, which first nested in 1996, now total 800 breeding pairs; and the first two pairs of cattle egrets bred in heronries on the Somerset Levels in 2010. The very first British nesting record of the little bittern occurred there that year and the first of the great white egret in 2012.

These birds are able to discover places to nest now that the conservation bodies have got around to creating reed beds and marshes out of former farmland and abandoned peat diggings on the so-called Avalon marshes in Somerset. It helps that many of the species are colonial nesters and can readily find mates once they arrive at a suitable location; this counters the 'Allee effect' whereby declining species continue to decrease if their remnants are too scattered for birds to meet partners. The spill-over of birds from Holland means that the Dutch may almost be said to be repairing the avian damage wrought by Vermuyden.

ENDNOTES

[i] C. Wilson, *The Dutch Republic* (London: Weidenfeld and Nicolson, 1968). Other sources for this chapter include Barbour, *Capitalism in Amsterdam*; G. E. Fussell, 'Low Countries' Influence on English Farming,' *English Historical Review*, **74** (1959), pp. 611–622; M. G. Hatvany, *Marshlands: Four Centuries of Environmental Change on the Shores of the St. Lawrence* (St.-Nicolas, Quebec, 2003); J. Purselove, *Taming the Flood: A History and Natural History of Rivers and Wetlands* (Oxford: Oxford University Press, 1989); A. W. Skempton *et al.*, *Biographical Dictionary of Civil Engineers in Great Britain and Ireland* (London: Thomas Telford, 2002, Vol. 1); and local histories of places in England subject to early drainage schemes, many of them involving Dutch engineers.

[ii] G. P. van de Ven (ed.), *Man-made Lowlands: History of Water Management and Land Reclamation in the Netherlands* (Utrecht: International

Commission on Irrigation and Drainage, 4th revised edn., 2004), p. 98 and passim.

[iii] Anon., 'The History of trawling,' *The Fish Trades Gazette*, 19 March 1921, p. 25.

[iv] A. W. Skempton, *Biographical Dictionary*, pp. 831–835.

[v] G. E. Fussell, 'Low Countries' Influence,' pp. 619–622.

[vi] C. T. Smith, 'Dutch peat digging and the origin of the Norfolk Broads,' *Geographical Journal*, **13** (1) (1966), pp. 69–72.

[vii] E. L. Jones, *Locating the Industrial Revolution* (Singapore: World Scientific), pp. 31–32, 38.

[viii] M. A. Knittl, 'The design for the initial drainage of the Great Level of the Fens,' *Agricultural History Review*, **55** (1) (2007), pp. 23–50.

[ix] C. Weaver, 'From private pursuit to public playground: The enclosure of Malvern Chase,' *Transactions of the Worcestershire Archaeological Society*, 3rd ser., **16** (1998), p. 211.

[x] L. E. Harris, *The Two Netherlanders: Humphrey Bradley, Cornelis Drebbel* (Leiden: Brill, 1961), p. 106.

[xi] Barbour, *Capitalism in Amsterdam*, pp. 124–125.

[xii] G. P. Van de Ven (ed.), *Man-made Lowlands*, p. 139.

[xiii] J. Turnbull, *The Scottish Glass Industry 1610–1750*, Society of Antiquaries of Scotland, Monograph Series, n.d.

[xiv] R. H. Fisher, *The Russian Fur Trade, 1550–1700* (University of California Publications in History, XXXI, 1943), p. 192. The fur trade was one of the many where the Dutch acted as middlemen.

[xv] O. P. Kennedy, 'The Lough Neagh Fishery,' *Newsletter of the Inland Waterways Association of Ireland*, **27** (3), August 2000.

[xvi] A. Coney, 'Fish, fowl and fen: landscape and economy on seventeenth century Martin Mere,' *Landscape History*, **14** (1992), pp. 51–64.

[xvii] A. E. Smith and R. K. Cornwallis, *The Birds of Lincolnshire* (Lincoln: Lincolnshire Naturalists Union, 1955), p. 14.

[xviii] For the possibly independent invention of decoys in England, see R. Lubbock, *Observations on the Fauna of Norfolk* (Norwich: Jarrold and Sons, 1847; 1879 version edited by Thomas Southwell), p. 145, n. 122.

[xix] R. Lubbock, *Fauna of Norfolk*. See also C. Dixon, *Lost and Vanishing Birds* (London: John McQueen, 1898); and M. Cocker and R. Mabey, *Birds Britannica* (London: Chatto & Windus, 2005).

[xx] R. S. R. Fitter, *The Ark in Our Midst* (London: The Country Book Club, 1961), pp. 218, 249.

[xxi] The pervasive effects of game preservation are indicated by Southwell's note (in Lubbock, *Fauna of Norfolk*, p. 85, n. 83) that all the young herons in a Norfolk heronry were killed by a gamekeeper to furnish his pheasants with maggots.

[xxii] E. A. Ennion, *Adventurers' Fen* (London: Methuen, 1942).
[xxiii] A. Bloom, *The Fens* (London: Robert Hale, 1953).
[xxiv] R. J. O'Connor and M. Shrubb, *Farming and Birds* (Cambridge: Cambridge University Press, 1986), pp. 96–97.
[xxv] A. H. Patterson, *Nature in Eastern Norfolk* (London: Methuen, 1905), p. 108.
[xxvi] A. C. Smith, *The Birds of Wiltshire* (London: Porter, 1887), p. 387; S. Marlow, *Winifred: A Wiltshire Working Girl* (Bradford-on-Avon: Ex Libris Press, 1993), pp. 24–25.
[xxvii] G. A. B. Dewar, *Wildlife in Hampshire Highlands* (London: J. M. Dent, 1899), pp. 132–133.
[xxviii] A. H. Patterson, *Nature*, pp. 70–71.
[xxix] R. Lubbock, *Fauna*, p. 132, 134.
[xxx] Protection of Birds Bill, Hansard, House of Lords Debates, 29 April 1954, Vol. 187, cc. 201–272.
[xxxi] J. Sheail, *Nature in Trust: The History of Nature Conservation in Britain* (Glasgow: Blackie, 1976), pp. 35–36.
[xxxii] R. Lubbock, *Fauna*, p. 96, n. 92.
[xxxiii] R. J. O'Connor and M. Shrubb, *Farming*, p. 95.
[xxxiv] C. A. Johns (revised by J. A. Owen), *British Birds in Their Haunts* (London: George Routledge and Sons, 1909 [first edition 1867], p. 282; A. H. Patterson, *Nature*, p. 249; J. E. Kelsall and P. W. Munn, *The Birds of Hampshire and the Isle of Wight* (London: Witherby, 1905), p. 334.
[xxxv] R. Lubbock, *Fauna*, pp. 171, n. 165, 172; E. R. Suffling, *Land of the Broads* (London: L. Upcott Gill, 1887 [first edition 1885]), pp. 135–136.
[xxxvi] C. A. Johns, *British Birds*, p. 282.
[xxxvii] A. C. Smith, *Wiltshire*, p. 287; J. C. Atkinson, *British Birds' Eggs and Nests* (London: George Routledge and Sons, 1870), p. 116.
[xxxviii] R. J. Cocker and M. Mabey, *Birds Britannica*, p. 234. The source is an individual without supporting documentation.
[xxxix] J. Copping, 'Top restaurants face a shortage of seagull eggs,' *Daily Telegraph*, 28 March 2009.
[xl] S. Moss, *Birds* (May, 2011), pp. 38–43.

EUROPEAN EXPANSION

10. EUROPE'S EXPANSION OVERSEAS

It is not by the importation of gold and silver that the discovery of America has enriched Europe.

Adam Smith, *The Wealth of Nations* (1776)

Humanity's earliest environmental impacts are known only indirectly, although the case has long since been put for a large-scale effect as early as the Palaeolithic age.[i] A more rewarding avenue for examination is the sequence of changes in agricultural systems, since farming is the main force transforming land use. Agricultural history is organised in different ways. One approach emphasises variations in the distribution of crops and livestock according to geology, topography or land use, while another concentrates on the ownership and occupation of land. Surprisingly few histories make their centrepiece the technical aspects of husbandry, that is to say the actual conversion of inputs into outputs. Fewest of all are treatments that focus on markets dealing in farm products or (in environmental histories) markets for products gathered from uncultivated areas. Yet the import of products collected from the wild paralleled agriculture in influencing Europe's trade and development and brought about change in global ecosystems. They were important in the centuries before manufactured products and synthetic materials became common. The outer world was treated as a giant open access resource.

Throughout much of prehistory, the most intensive and productive farming lay in the great river valleys from the Nile eastwards.[ii] Valleys were the seats of irrigated agriculture and the earliest highly organised societies, whose rulers could collect taxes and recruit armies from dense populations of peasant farmers. In the warm river valleys of the Middle East and Asia, the supply of grain permitted cities to grow. Colder climes developed their rain-fed agricultures more slowly.

What altered this pattern were huge diffusions of crop species via human migration and colonial movements. Crops were transferred in both directions between old and new settled regions. Notable examples included the introduction to Africa of Asian food plants such as yams, taro and bananas, which began about the start of the Christian era, though it took until about A.D. 500 for some crops to cross from east to west Africa. Another movement is termed the 'Arab Agricultural Revolution' and is attributed to the spread of Islam in the eighth century of the Christian era. This diffusion supposedly brought 16 food crops (plus cotton) from as far away as India, westwards into the Middle East, North Africa and southern Europe. The crops plausibly filled little-occupied niches in the Middle East and Mediterranean basin, which were otherwise hard to exploit because of hot summers and poor soils. Although recent research has shown that some of the crops were already present in the receiving areas, and Muslim farming practices were not fundamentally different from earlier ones around the Mediterranean, there were some displays of novelty.[iii] Another significant introduction was the transfer of early-ripening and drought-tolerant Champa rice from Vietnam to China in the twelfth century A.D.

Most far-reaching of all was the 'Columbian Exchange'. Astonishingly soon after they had entered the New World, Europeans redistributed American crops to Africa and as far away as China. The extent of diffusion may be gauged from the fact that between the start of the Christian era and A.D. 1500, only three new crops entered China, but during the sixteenth and seventeenth centuries, five major dry-land crops were brought in. Transporting plant specimens was relatively easy; it increased productivity in the recipient areas without much embroiling them in international trade. Technology transfers raised output *within* individual continents; they did not bring about a convergence of prices in the different continents, which is the test of significance in trade theory, yet expressed a deeper significance through supporting population growth.[iv] Some areas where Europeans settled were very soon able to despatch surpluses homewards. Barbier gives as instances the Azores, Canaries and Madeiras, which were among the first and closest of Europe's colonies.[v] Their ecosystems were almost totally altered by transplanted agricultural systems.

Trading with distant and virtually unknown lands meant long voyages with results that were highly uncertain, the more so because of competition among the Europeans. National pride was at stake, not to mention the urge to spread Christianity. These motives inspired exploration at the behest of crowned heads who, despite commanding considerable resources, were not invariably generous in backing the adventurers they sponsored. But their impact could be profound, as when Henry the Navigator founded a college in the fifteenth century to systematise what could be gleaned of the extra-European world and equip his sailors with useful knowledge.

The usual means of spreading risk was the joint-stock company, incorporated under a charter from the ruler of the home country. Capital in joint-stock form and a monopoly of trade with a given region were not quite enough to still apprehensions about the risk and hence companies of merchants were granted additional powers that would ordinarily have been those of government.[vi] Cheyney gives a long list of chartered companies founded between 1554 and 1698.[vii] The Dutch seem to have been the most ambitious, founding a 'Company for Distant Lands' in 1594, an ambient claim if there ever was one. Normally, companies were named after the area in which they were licenced to operate. The very first had been the English Muscovy Company of 1554, which shows that more than American and African ventures were in mind. European resource expansion was to the east as well as the west, something easily forgotten amidst the geopolitical drama of the Atlantic trade.

The purposes behind the ascending volume of exploration changed as information about overseas prospects filtered back. Few endeavours had a single aim; and while trade and colonisation, not to mention the hope of windfall resources, figured largely, there were scientific aims as well. Linnaeus's northward journeys can be cast as joint resource appraisal and scientific investigation although the commercial results were scant.[viii] Well-to-do amateur scientists in England were prominent among the sponsors of botanical and other exploration in the American colonies. Much correspondence survives to record the work of enquiring individuals, which presaged quasi-official endeavours such as those personally involving Sir Joseph Banks, as well as other voyages under his aegis as President of the Royal Society. In the nineteenth century, exploration pushed out in all directions and its mixture of motives is particularly well illustrated by the work of the Illinois naturalist and scientist, Robert Kennicott. He led the Western Union Telegraph expedition to Russian America and it was his resource appraisal that persuaded the United States to buy the area and rebrand it as Alaska. The name of Kennecott Copper is no coincidence, despite the misspelling of his name.

The Columbian Exchange as a technology transfer grandly increased the portfolio of species available to farmers; it was the biggest single positive shock brought about by redistributing the world's biological potential. American crops came into Europe, Africa and China. Meanwhile, Old World crops, such as sugar, cotton, rice, indigo, wheat, grapes and olives, besides cattle and horses, reached the Americas. The prospects were brightest for farming and food exports in the so-called neo-Europes, the temperate zone regions of the Americas and Australasia so like parts of Europe that it has occasionally been imagined they were pre-adapted to receive its settlers. Similarly, the tropics could be modified and exploited but Europeans never found them as congenial or medically safe. European settlement or not, the effects of new crops were formative in all the regions reached. But the shipping technologies of Victorian times were needed before bulk intercontinental trade in foodstuffs could develop.

History was asymmetric. During the three centuries after Columbus, the human population of the earth rose almost seven times more than it had during the preceding three centuries. Europe's share of this growth was disproportionate, with an unprecedented surge to follow in the nineteenth century, when there was a migration to the Americas of 40 million people — no wonder Walter Bagehot called them, 'the conquering *swarm*'. World population growth as a whole was accompanied by massive conversions of land to growing crops. The arable area in Japan rose by 200% between 1200 and 1800, in China by 150% between 1400 and 1760–1770 and in India by 71% between 1600 and 1900. At a time when Asia was becoming more densely settled, population per square mile in the vast joint region of Western Europe plus the New World moved in the opposite direction, plunging from 26.7 in 1500 to 9.0 in 1800. For people of European origin in the Americas and Australasia, more numerous though they were, came an unimaginably large accession of farmland per head. The thin spread of colonists nevertheless proved entirely capable of creaming off the most readily accessible natural resources.

From the viewpoint of northwestern Europe, and above all of Britain, this resource base constituted a vast expansion of 'ghost acreage'.[ix] The concept refers to the notional extra area of land (there was also marine ghost acreage) that would have been needed at home to produce resources equivalent to those imported. Europe would have had to swell in size — out of the question in physical terms, obviously, but possible in the shape of virtual growth through trade and colonisation. From the viewpoint of the outer world, the expansion meant the alteration of whole landscapes, with multitudinous ecological changes.

Many of the changes were individually minor; the transformation of the earth came about through adding them to the spread and intensification of agriculture, so that the world became a series of syncretic ecosystems with no region untouched. The spread of disease organisms is attributed to 'microbial unification'. Plants were exchanged among the regions outside Europe. In the early seventeenth century, Clusius persuaded the VOC (the Dutch East India Company) to collect plants from Southeast Asia. The Company was eager to supply its employees in the East with better medicines than it was already acquiring from Arabia. Supplies from Ceylon were replaced in turn by medicinal plants from Malabar in India. The 12 volumes of *Hortus Malabaricus*, printed in Amsterdam between 1678 and 1693, contained 700 illustrations of medicinal plants, with explanations of their use.[x]

Europe itself received desirable and undesirable species from throughout the animal and plant kingdoms: hundreds of new trees came in but also dry rot, which was imported to eighteenth-century Britain in timber from the Himalayas.[xi] Other introductions were ornamental, like buddleia, which had been brought from the Caribbean about 1730. The variety of buddleia called *davidii* came from China to Kew Gardens in the 1890s and now occurs widely in gardens, besides spreading so vigorously on wasteland as still to be termed an invasive species. Buddleia is known as the 'butterfly bush' and has revolutionised England's garden butterfly fauna.

Wherever Europeans went to settle, introductions from 'home' became fashionable. As a result, New Zealand has been called a 'witless menagerie'. Australia had its acclimatisation societies, and the ultimately unwelcome introduction of the rabbit, fox and other pests. Introductions to North America were sometimes less successful. Foxhounds were taken to India where the packs were used by army officers to hunt jackals — testimony, if one be needed, that hunting was intrinsically little to do with controlling foxes but concerned elite bonding and military riding practice.[xii]

Europe's overseas expansion did not rest with the colonies and trading stations of its initial phase. Resource frontiers of every type continued to roll outwards. As each was worked to the point where the search costs rose steeply, exploitation moved to another frontier or some fresh commodity. The logging frontiers of North America were major examples, as were the whaling grounds of the south Atlantic and far Pacific. The cornucopia of natural bounty, endless forests and grasslands, rivers teeming with fish, coastlines rich with seabird and seal colonies, seas full of whales, encouraged prodigality and led men to believe that total supplies could never run dry. But they did. The outcomes were more intensive exploitation at rising marginal cost, ever more exacting searches for fresh pastures, and the substitution of farming where there had been only hunting. This is evident in the massive increase of fish farming where fish had initially been caught by men playing nothing more than lucky dip.

Substitutions of various types were late-stage results of Europe's expansion. The commercial production of rabbit furs in English warrens was a cheaper replacement for imports of Russian squirrel and other mammal pelts. Subsequently, the warreners were outcompeted by the importers of North American beaver fur (British beavers having become extinct, though there had never been enough to support much of an industry). But while domestically produced rabbit fur was inferior to beaver for making hats, the growth of the market was such that the warrens remained in business for some time. The trade-off between domestic honey and imported sugar operated in a similar way: sugar imports were large but those who could keep hives in the garden or on common land continued to find it cheaper to use their own honey. It was at the elite end of the food market that the substitution of imports, or imported species, for domestic products was most decisive, as when feasting on swans or peacocks gave way to the breeding of turkeys — American birds despite their name.

The pressure of demand on domestic English production soon had measurable consequences. Under-age fish that would normally have been left to mature were now caught for the London fishmongers. Although salmon was still a common species in the Thames, it was profitable as early as 1714 to catch them far away in the River Severn for sale in London, 'of unsizeable lengths and at unseasonable times'.[xiii] Salmon from the Tweed had to be salted, dried, boiled or pickled in vinegar, and were several weeks on the journey to London, until about 1790 'one Marshall' succeeded in sending them 'fresh from the water' on six horses.[xiv] They were packed in ice which was collected in the winter and stored in ice houses until well into the summer. Because trade within England was free trade, it was not recorded in the systematic fashion that tax or customs' records of overseas commerce involved and the precise scale is unknown.

The import of foodstuffs from outside Europe became truly substantial in the nineteenth century. At that period, when shipping technologies and capacity improved, the shock of cheap grain was felt severely by European cereal growers. Arable farming in Europe was beleaguered and where producers were not protected by tariffs or subsidies, their land became tumbledown. Those who could do so staged a retreat into specialist livestock production, creating different types of landscape and affecting wildlife in new ways. Livestock farming's greater natural protection of distance was eroded in turn by refrigeration. This made it possible to ship meat and butter from the Antipodes; only liquid milk production remained unscathed. The other face of Europe's late nineteenth-century depression was the so-called Pioneer Agricultural Explosion, when lands outside Europe were cleared simultaneously, vegetation burned on an unprecedented scale and carbon dioxide poured into the atmosphere.

ENDNOTES

[i] R. F. Heizer, 'Primitive Man as an Ecologic Factor,' *Kroeber Anthropological Society Papers*, **13** (1955), pp. 1–31.

[ii] E. Jones, *The Record of Global Economic Development* (Cheltenham: Edward Elgar, 2000), pp. 48–63.

[iii] M. Decker, 'Plants and Progress: Rethinking the Arab Agricultural Revolution,' *Journal of World History*, **20** (2009), pp. 187–206.

[iv] E. Jones, *The Record*, pp. 53–55.

[v] E. B. Barbier, *Scarcity and Frontiers* (Cambridge: Cambridge University Press, 2011), p. 242.

[vi] Another means of dispersing the risks of resource exploitation was taking shares in whaling vessels instead of owning them outright.

[vii] E. Potts Cheyney, *European Background of American History 1300–1600* (New York: Collier Books, 1961), pp. 86–88.

[viii] M. Asberg and W. T. Stearn, 'Linnaeus's Oland and Gotland Journey 1741,' *Biological Journal of the Linnaean Society*, **5** (1973), p. 3.

[ix] E. L. Jones, 'The Discoveries and ghost acreage', Chapter 4 in *The European Miracle* (Cambridge: Cambridge University Press, 3rd edn., 2003).

[x] See www.botanicus.org and www.hortusmalabaricus.net.

[xi] C. Dodd, 'The dry rot detective,' *Heritage Today*, **31** (1995), p. 58. Approximately 20 tree species had been introduced to England by 1600, another 32 by 1700, 63 more by 1800 and 300 during the nineteenth century. (Diagram prepared by the late A. F. Mitchell, Alice Holt Forest Research Station). On introductions of birds and animals to Britain, see R. S. R. Fitter, *The Ark in Our Midst* (London: Country Book Club, 1961) and C. Lever, *The Naturalised Animals of the British Isles* (London: Paladin Books, 1979). On introductions world-wide, see J. L. Long, *Introduced Birds of the World* (Sydney: A. H. and A. W. Reed, 1981).

[xii] W. Dalrymple, *City of Djinns: A Year in Delhi* (London: Flamingo, 1994), p. 77.

[xiii] P. Hurle, *Upton: Portrait of a Severnside Town* (Chichester: Phillimore, 1979), p. 11.

[xiv] C. L. Cutting, in Anon. (ed.), *Getting the Most out of Food* (London: Van den Berghs, 1967), p. 51.

11. EUROPE'S DISTANT REACH

> Nor could it be imagined that this Wilderness [of Colonial America] should turn a mart for Merchants in so short a space, Holland, Spain, France and Portugal coming hither for trade.
>
> Edward Johnson (1653)

European voyagers were at once impressed by the bird-life of the Atlantic and the New World. Columbus reported to King Ferdinand and Queen Isabella that, on the islands of the Caribbean, he had found himself surrounded by 'birds of a thousand sorts.' Island groups were named the Canaries, the Azores (from the Portuguese for hawks), Brazil was referred to as 'Parrot-land' and several islands in the Gulf of St. Lawrence and off Newfoundland and Labrador were called Iles des Oiseaux and Iles des Margaux ('Magpie Islands', after the black and white great auks).[i] There are ornithological records for the Atlantic islands as early as those by Fructuoso (1522–1591). Tropical villages of the East and West Indies were found to be overrun with pets and sailors of every European nation returned with monkeys and parrots. Soon, there were pet shops for the home population and zoos where the public might view the rarer or less domesticated species. Canaries became the most popular household songsters and, by about 1500, all the well-to-do in Portugal were said to be keeping them. Genoese and Tuscan ships began to visit the Canary Islands to buy canaries and the species began to appear in captivity in other parts of Europe. Rather than rely on the trapping of wild birds, by 1580 Spanish merchants were persuading peasants and small craftsmen to 'farm' canaries in numbers.

The uses of birds and other wildlife found in the new lands were typically greedy and often cruel; some representative detail will be provided here. The sheer numbers to be found disarmed all thought of restraint and explorers' accounts glory in the easy kill. On an island off North Carolina, members of Raleigh's expedition discharged their 'harquebus-shot' into a great flock of either or both whooping and sandhill cranes.

Captain John Smith recorded that 148 fowls were killed with 3 shots in Chesapeake Bay, where the wings of the wildfowl were described as 'like a storm coming over the water.' Dutchmen who settled in Delaware called their town Zwaanendael (valley of swans) and said the flocks obscured the sun.

When the Humboldt Current off the west coast of South America was reached, the richest zone for marine life had been found. Cormorants, pelicans and gannets flighted every morning from the offshore islands to feed on the shoals of surface-swimming fish. Their cries and booming wings created a concussion in the air. Protein-starved crews of tiny ships far from home were overawed by the abundance, there and elsewhere. They could not be expected to stay their hands as they fell on the unafraid hosts of seals, turtles and penguins. On his first trip to Newfoundland in 1534, Cartier's crews filled two boats full of great auks in less than half an hour. In 1578, Drake's men killed over 3,000 Magellanic penguins in one day on the island of Santa Magdalena.[ii] Nine years later, Cavendish called there again and 'powdred three tunnes of Penguins for the victualing of his ships.' Five years after that, John Davis landed a party which killed 14,000 nesting penguins and salted them down. Seabird colonies along the ocean routes functioned as free provisioning stations and it seemed as though supplies could never run out. 'Today boil'd shags and Penguins in the Coppers for the Ships Company's Dinner', logged the first lieutenant of the *Resolution* on Cook's voyage of 1775.

One species was driven to extinction by all this attention: an Arctic bird, the great auk. Its numbers dwindled in the late eighteenth century, and when the last two birds were killed in 1844, the surprise is that they had lasted so long. Thereafter, the price of specimens and eggs rose astronomically: an egg that would have cost a collector 12 shillings in 1819 was 525 times dearer by 1934. Ironically, the final sighting (and killing) of the great auk came only two years after the emperor penguin of Antarctica had first been identified — standing around as if waiting to be slaughtered by hungry seamen, just as the auks had seemed to do in the Arctic. Emperor penguins could weigh up to 60 or 70 lb. In the nineteenth century the carcases of these and other penguins were burned in South Georgia and the Falklands to melt the oil out of seal carcases. There was nothing new in seamen burning up natural resources in this way, literally and figuratively. In the early seventeenth century the English fishing fleet off Iceland had carried long-winged 'hawks' and sixteen or twenty falconers for provisioning the boats with sea-fowl.[iii]

In 1603, the crew of a galleon stumbled on the existence of the cahow, a gadfly-petrel, and promptly dried and stored thousands of it for food.[iv] Settlers sent to the eastern end of Bermuda during a famine so stuffed themselves with cahow flesh that some died from indigestion. By 1616, and again in 1621, the governor of Bermuda was trying to save the stock, as he also tried to save the green turtle, but there was no sign he succeeded. The cahow was luckier than the great auk — it appeared to vanish from the world under the ferocity of the onslaught. From the 1620s, the cahow was deemed to be extinct, until singletons were identified in the 1930s. The nesting grounds — of a world population fewer than 100 — were found in 1951. Early settlers in the West Indies almost wiped out an allied gadfly-petrel, the diablotin, by netting it at the burrows. On Guadeloupe, that species was indeed exterminated.

Far to the south, Matthew Flinders estimated in 1798 that the population of another petrel, the mutton-bird or short-tailed shearwater, near Three Hummocks Island was at least 100 million. Montgomery estimated at least 2.6 million on four islets in the Furneaux group alone. As with most numbers cited in the early literature, the figures are not to be taken literally. It is enough, more than enough, that they were prodigious, yet by the 1890s had been so reduced for feeding the sealers and by commercial exploitation that an appeal had to be made to the legislature to protect them.[v]

An enormous 'harvest without a seedtime' was thus gathered from the profusion of oceanic species. Just how high marine productivity was before European exploitation got under way may be gauged from an estimate that the world population of whales ate 100 million tons of krill per annum, while other marine creatures may have eaten an equivalent weight. Over 600,000 seal skins were imported to England in the mid-nineteenth century.[vi] Yet, despite the massacres of Victorian times, the seals still sweeping up and down the Pacific in the 1950s consumed an estimated 1.5 million tons of fish and squid every year, more than the entire catch of the United States' North Pacific fisheries.[vii]

Seamen and settlers pressed hard on sources of protein. Island species were particularly vulnerable and were reduced or rendered extinct by direct exploitation or the secondary effects of cats and rats released or escaped from passing ships. Palatable turtles and tortoises were hit especially hard. During the eighteenth and nineteenth centuries, seals were so hunted all around the world's shores that many unimaginably large populations collapsed, and by the end of the nineteenth century, several species were thought to be extinct. This was too pessimistic, although not by a wide margin.

Once hunting pressure abated, there were some remarkable recoveries, notably among the larger whales, yet the disequilibrium brought about in the marine life of distant oceans was strikingly different from the effects of conservative harvesting practices within Europe itself. The range of sea birds taken in Scotland and Faroe probably remained stable for thousands of years, though the author of a close study of fowling methods wisely notes that an absence of documentation about fluctuations in their populations should not really be taken as evidence of the absence of fluctuations.[viii] The great auk had gone, but the fulmar, he adds, had started nesting in the area only within the previous 100 or 150 years, after beginning to expand around Iceland in the mid-eighteenth century.

The tragedy of the great auk and the negative human impact on seabirds are well known. The positive side of the story is less familiar. It stems from the dumping of waste by fishing boats and trawlers, and first probably by whalers. In the opinion of Fisher and Lockley, the effect on the fulmar was unprecedented and unmatched: 'the biggest revolution in the numbers of any widely-spread sea-bird (or any bird)'.[ix] More recently, the number of vessels fishing from European ports has declined. Gulls that formerly concentrated at the ports to feed on fish waste have moved to inland rubbish dumps, and they nest on and foul buildings in towns where other food waste is to be found, well away from the sea. Waste produced by human society is a large subsidy to the gull population, as it was to the fulmar.

THE PULL OF THE MARKET

Before the mass production of synthetic materials, consumer goods often consisted of, or contained, biological items. And by the nineteenth century, many of the world's ecosystems were being re-ordered to supply such things to manufacturers in north-west Europe. The wealth and drawing power of the industrial countries secured a supply of raw materials far beyond what could be found in their immediate surroundings. Purchasing power in the developed world was an irresistible draw. 'There is no gold mine of any importance, but there is more gold in England than in all other countries,' wrote an American, Ralph Waldo Emerson, in the nineteenth century, 'it is too far north for the culture of the vine, but the wines of all countries are in its docks; and oranges and pine-apples are as cheap in London as in the

Mediterranean'.[x] London acted like a vacuum cleaner, distant regions were subordinated to it and some were stripped almost bare of items that could be gathered from the wild. Search and collecting costs rose steeply but catchment areas were extended one after another, until the world turned into one giant catchment.

For the fur trade this had begun well before the nineteenth century. Furs and timber products were among the earliest 'merchantable commodities' sent back to repay American settlers' debts to their financial backers. New England may have been founded for religious reasons, as a 'City on a Hill', but this meant neither that the Puritans owned enough capital to finance colonisation themselves nor that they were exempt from commercial urges. Almost at once some of them began to speculate in land. In 1621, they sent home the *Fortune* (55 tons) carrying a mere two barrels of furs but also 'laden with good clapboard as full as she could stow'.[xi] New England became important to Old England as a source of naval stores, especially timber from what became a white pine frontier pressing inland as fast as Indian resistance and difficulties with transport permitted.

The marketing of wildlife materials and the trades handling them are not easy to fathom. Data exist for the shipment of furs by the Hudson's Bay Company and the whaling industry also generated much numerical information. More often only broken runs of figures survive with which to gauge the scale of activities at home or abroad, like those Pearsall quotes for sales in the nineteenth century at the biggest British fur mart, the Candlemas Fair at Dumfries.[xii] Proximity to markets was what was important to processors, although the siting of processing industries is misleadingly explained by their physical resource advantages. As a case in point, the cluster of twenty great dyeing firms at Weissenfels, Germany, engaged in colouring Russian furs bought at nearby Leipzig fair (reputedly the world's largest fur mart), is explained by the special presence of clays and salts used in dyeing.[xiii] But this is *post hoc ergo propter hoc* and Agnes Laut is sceptical that geology always explains commercial location: 'the supposition I again doubt', she writes, 'for American firms are doing the same dyeing today and doing it well'.[xiv] Processing used chemicals from the local soils but this does not show they were the primary factors. They could have been imported if need be.

In the early seventeenth century, Amsterdam ruled international commerce, reaching out to a whaling station on Spitsbergen, with one thousand hands in the season; to New Amsterdam (New York) and the Hudson Valley, where Dutchmen settled; and across the world for spices from Java.

The Dutch bought the skins of fur-bearing mammals in Russia and sent them to Western Europe. One fur trade after another was opened up.[xv] The ability to shift from a given source of supply to another, and from one species of fur-bearing mammal to the next, disguised the end-game from those playing it, though they would have continued anyhow until costs rose to stifle all profit. The threat did not hoist warning signals as long as any new region or fresh species remained available. Between the Middle Ages and the nineteenth century, in the endless forests of Russia and America, no end seemed conceivable let alone in sight.

English-language works on world resources dwell mainly on the American contribution. They do not fairly reflect the time when Russia was the greater reservoir, exporting furs to Europe and providing the bulk of beeswax, tree honey, tallow, hides, seal oil, sturgeon, flax and hemp, salt and tar.[xvi] Ever since the Middle Ages Western Europe had drawn resources from Russia, from Eastern Europe as a whole, and from Scandinavia. (Karl Appuhn points to a lesser known trade whereby droves of cattle were brought to Italy after being reared on the Hungarian Plain).[xvii] With brio scarcely equalled in the West of North America, Russian settlement pressed eastwards — and earlier too.

At the medieval height, Russia had been exporting 400,000 to 500,000 squirrel skins per annum but in the early fifteenth century imports into London declined and rabbit skins from English warrens took over. The Russians responded by moving their hunting grounds to the east of the Urals, distances that did little to moderate prices. During the seventeenth and eighteenth centuries, over 2 million settlers entered the wooded steppes and the open steppes, and 400,000 entered Siberia. The Bering Straits were reached by 1649 and the fur trade further expanded.

Russia looked the other way for a market and in 1727 signed the Treaty of Kiakhta with China, whereby furs were exchanged for silk. But by the late eighteenth century, Siberian fur supplies were shrinking and the Russians were on quests for new species (such as the sea otter) that took them to the Aleutians and Russian America. Demand was still barely satisfied and they even re-exported American beaver to China![xviii] In the nineteenth century, the trade tended to peter out and its decline by the 1860s was one consideration in the country's willingness to sell Russian America (Alaska) to the United States.

In mid-Victorian times, the Hudson's Bay Company was trading in the skins, pelts or furs of 16 species. The English import of squirrel pelts was over 2 million per annum.[xix] But land-bearing mammals were becoming harder to find, and in the late eighteenth century, sea otters were sought; when they decreased in turn — 250,000 having been killed between 1750 and 1790 — seal skins were the next target. In other regions, the replacement of forest by farmland was driving back the larger animals. Unlike birds that benefited from deforestation, among the mammals it was only rabbits (and in the United States, opossums) that extended their range and numbers.

European expansion may have begun out of 'no necessity' but without the resources of its 'ghost acreage' Europe could scarcely have achieved its standard of living. Its population grew faster than that of other world regions, breaking away, as McEvedy and Jones put it, 'into a class of its own with a 19th-century gain of 115%'.[xx] Not surprisingly, then, Victorian times saw the peak of Europe's onslaught on the world's wildlife. As Crosby declares, the replacement of much of the flora and fauna of the Americas and Australasia by the farm crops and animals, pests and weeds brought by Europeans had thunderous ecological consequences.[xxi] Despite losses and reductions of range among indigenous species, the introductions increased biological variety overall, as well as presumably raising the biomass. From the economic point of view, Crosby notes, this was a second miracle of the loaves and fishes. He points out just how many people in the world have come to depend on food from the Neo-Europes.

Within Europe, communities tried to ration resources among themselves. The governance of common fields and common grazing was subject to intricate rules designed to prevent short-term abuse — and effectively so.[xxii] Control, or self-control, extended to coastal waters. In eighteenth-century Cornwall, a system of 'stems' parcelled out the use of small patches of offshore water among the seine fishermen, one day at a time.[xxiii] Over-use was frowned on, and resources were kept in the hands of local communities, with 'strangers' excluded. As populations rose, however, no purely defensive system could prevent heavier and heavier exploitation or frustrate every land grab.

At the same time, efforts were made to raise productivity. Following political and organisational struggles within country after country, rigid agrarian systems were reformed and best practice husbandry diffused. Where fresh land could be found, it was put under cultivation. In Sweden, there was an internal frontier movement throughout the eighteenth century; farmers moved north, clearing boulders as they went. But in long-settled areas, the prospects of expanding the cultivated area were limited and the additional land taken in — hitherto shunned by the plough — was on average of inferior fertility. Some developments merely shifted the environmental costs to fresh parts of the continent. Appuhn argues that by importing droves of cattle from the Hungarian Plain, Western Europe avoided the costs of converting arable land into pasture to raise beef.[xxiv]

Certainly farming practices were improved and new crops were introduced, some of them originating outside Europe. Fertilisers were so scarce that the agricultural chemist, Liebig, accused Britain — perfidious Albion — of scouring the battlefields of continental Europe to make bone-dust fertiliser from the skeletons of dead soldiers. Means were found to improve the texture and chemical composition of soils, notably by liming, though this became possible on a grand scale only when coal to burn the lime was extensively mined and transported by rail. These activities were not invariably the products of a self-sufficient Europe because some of them relied on imported materials. Nor could they have raised productivity enough to satisfy the needs of the fast-growing urban populations of nineteenth century Europe. For that, resources from outside the continent were indispensable: trade, more trade, was the key. Sources of imported fertility like Peruvian guano made a real difference.

THE LONDON MARKET

During the eighteenth century, and overwhelmingly during the nineteenth, London took over from Amsterdam as the metropolis of international commerce, though the entirety of northwest Europe is often described as 'metropolitan'. This meant a vast extension of England's internal trade, which had of old centred on London, whose population grew from 1 million in 1801 to 4.5 million by 1901. For centuries, London had reached out for foodstuffs around the coastline and received cattle and sheep driven overland from Wales. Now it reached much, much further.

At first, the capital's grasp was not as good as its reach and its early dominance is easy to exaggerate. In 1640, London contained barely 7% of the population of England and Wales. Much of the Midlands could not yet send loads of grain to London for lack of navigable rivers flowing in the right direction: Midlands farmers were 'lock'd up in the Inland.' But by the first half of the eighteenth century, London was a spider at the centre of a great web; its growth, coupled with improvements in transport and communications, ensured that only 'islands' of countryside remained outside its supply zone. The number of mouths to be fed in the capital was becoming prodigious.

To illustrate the effects, let us consider London's nineteenth-century consumption (and redistribution) of poultry and all species of wild birds:

> 'When we consider that fifty years since the sale of game was illegal', remarked an article in *Leisure Hour* in 1889, 'what a "development of commerce" have we here! And yet how small a matter it is after all! Add all the quantities together and it will be found that reckoning them all as equal, from the lark to the turkey, all at so many head, we have not allowed two birds a piece per year for the people of London. But how many of the Londoners we may be reminded taste not bird-flesh from year's end to year's end. Full well we know it. But take "the upper ten thousand," not as a figure of speech but as a figure of arithmetic, and allow each of the ten thousand a bird a day and you have accounted for 3,650,000 of the mighty flock!'

This is a mildly contorted way of acknowledging the unequal distribution of income and explaining how the throughput of Leadenhall market, the greatest market for poultry, could ever have been eaten. Like so many historical statistics, the number is not to be taken literally — the author of the original article admitted that, 'Leadenhall likes not statistics' — yet the true total was undoubtedly large.

The variety was large, too, and is worth enumerating:

> 'game innumerable, all hanging dead in plumes; and venison, skinned and unskinned, such as this market distributes in London alone to the tune of 350 tons a year. Where does the game come from? Scandinavia and Russia, Germany and Italy, Manitoba and Wisconsin all contribute. Even "the quails of the desert" come to Leadenhall. The grouse come from Yorkshire and the Highlands of Scotland; the pheasants and partridges from Norfolk and Suffolk; the teal, the widgeon, and the wild fowl from Lincolnshire and Cambridgeshire and' — be it noted — 'the lowlands of Holland.'

Leadenhall might not like statistics but Victorian taste did run to encyclopaedic descriptions amply laced with numbers. Hence it was felt worth reporting that:

> . . . a well-known salesman estimated that there are supplied in one year 100,000 grouse, 125,000 partridges, 70,000 pheasants, 80,000 snipe, 150,000 Irish plover, 30,000 Egyptian quail, 70,000 widgeon, 30,000 teal, 200,000 wild duck, 150,000 small wild birds, and 400,000 larks. And to this he added 400,000 pigeons, mostly from France; 100,000 geese, mostly

> from Holland; 350,000 ducks, a good many from Buckinghamshire; 104,000 turkeys, mostly from East Anglia; 100,000 hares, 1,300,000 rabbits, and 2,000,000 domestic fowls, mostly from Surrey and Sussex.

What retailers would call the 'offer' at Leadenhall included numbers of species, 'all tempting the amateur as ornaments for the back yard, and all meaning roast or boiled in the immediate future'. In other words, they were bought live and fattened for the table at home. There followed a miscellany of 'unprofitable pets', miserably huddled together but at least not destined to be eaten. They were 'hawks and canaries, larks and linnets, parrots and owls, hedgehogs, goldfish, foxes, water snails for the aquarium.' From all this it is apparent just how capacious was London's maw, although since the capital was also a grand processing and redistributing centre not everything that arrived was kept for local, or even English, consumption. In the mid-nineteenth century, a sizeable proportion of the English take of lamprey, which varied between 100,000 to 800,000 per annum, was exported to Hamburg, Danzig and the Netherlands, either for food or to be used as livebait.[xxv]

The demand for commodities ordinarily produced or able to be gathered from the countryside became so large it was hard to meet from English sources alone. Duck from the Netherlands appeared not only in London but in Norwich, a city lying right next to the prime English sources of waterfowl. Reaching out for overseas supplies became a feature of Victorian times, when a use was found for every biological item — body parts from throughout the animal kingdom and many an obscure plant.

The catalogue of these items is so exhaustive it would be wearisome to repeat but take just two examples from the medical field. First, leeches were collected for sale to doctors but supplies proved insufficient, and by the mid-nineteenth century, the main source had become the continent, especially France, Germany and Portugal.[xxvi] French stocks became so endangered that a decree was passed against exportation. Leeches may have been falling out of use in English medicine but even so a total of 15 million or 16 million were applied to hapless patients every year and a capital of £ 30,000 was employed in the trade. Secondly, herbs: at the start of the twentieth-century England possessed several *materia medica* farms of its own but for many years had been obtaining further supplies from central Europe, especially Germany and Austro-Hungary.[xxvii] The English herb acreage was declining in the face of this competition.

BELIEFS AND CONSEQUENCES

The limiting case of biological exploitation is extinction. If eliminating wildlife does not 'matter' in some deep evolutionary sense, as has been proposed, wanton destruction would still seem pointless.[xxviii] The dodo, the great auk, the passenger pigeon — no further lament is needed. Nevertheless, natural scientists' calculations of the rate of extinction are extraordinarily devoid of documentation. For instance, Professor Michael Benton claims that 130 species were driven to extinction by hunting between 1500 and 2000.[xxix] No evidence is presented but the rate of loss is said to have been much above the background rate.

The problem is that the background rate cannot be known, because the denominator — the initial world population of fauna and flora — is unknown. As Briscoe and Aldersey-Williams say bluntly in their discussion of the calculations proffered in the literature, 'we don't really know what we are talking about'.[xxx] There is no agreement about the number of species that exists now, let alone in the past. The rates of loss are usually predicted by a mathematical model based on habitat loss, since this is said to be more easily 'measured' than species loss. The most publicised estimates predict that habitat loss will ensure that between 18% and 35% of all species will have become extinct by 2050.[xxxi] Thanks to a conceptual error that went unnoticed for decades, the estimate may be two or two-and-a-half times above the true figure! This is astonishing, given the strident claims made for the superiority of arguing by numbers: no numbers can be involved here since none exist. Tractability trumps science, let alone history. The cost of (pseudo-) precision is the chance of calamitous errors that would be more easily spotted in language than in arithmetic. Fortunately research may be more closely scrutinised now that countries are being held to their biodiversity treaty obligations.

The catalogue of historical extinctions attributed to Westerners distorts the record badly. Sometimes, the losses are exaggerated as part of a political agenda, as when the author of an anti-free trade article asserts that, because of the economic growth of the previous 50 years, 'a quarter of the globe's bird species has been rendered extinct'.[xxxii] (Even if this were not absurd, no account is taken of the millions of children that economic growth has saved from starvation, disease and death). The insistence on the loss of 25% of all avian species appears and reappears, even occurring in *Science*.[xxxiii] But it cannot refer only to the past 50 years and completely misleads unless it is added that the prehistoric inhabitants of Pacific islands were the ones who succeeded in wiping out 20% of all the species of birds on earth.[xxxiv]

That effect instantly shrinks the proportion that could have fallen victim to the usual suspects, the Europeans. Greenway calculated in 1967 that 44 full species of birds had become extinct since the mid-seventeenth century. Views differ as to how many species there actually are and 44 is less alarming when one realises that it was a mere one-sixth of the discrepancy in the estimates![xxxv] In any event, some would probably have succumbed in the natural course of extinction and a few species have since been rediscovered. Greenway himself was aware that, because his work was 'limited to discussion of the discrete populations that have disappeared during the past 280 years, the geographical perspective is distorted'.[xxxvi]

Since the deed is done, attributing blame for the effects of Europe's expansion may seem beside the point. Yet European colonialism and two Western ideologies continue to be indicted: Christianity and capitalism. As to Christianity, Hancock pointed to incorrect readings of the Bible by the most influential environmental critic, Lynn White, Jr., as well as his failure to note that a drive for 'man's mastery over nature' began thousands of years earlier.[xxxvii] Hancock also criticised White's assertion that pagan man, 'respected the feelings of natural objects.' Was pagan man doing this (insofar as an object can have feelings), asked Hancock, when he set fire to forest and scrub? 'He was an inveterate incendiary'.

Lynn White's paper, 'The Historical Roots of our Ecologic Crisis', is the source most cited as blaming environmental damage on the Christian belief that nature was created for man's use. Refutations have been powerless in the face of this reading; it is what many wish to believe. The supposed novelty in Christianity was its abolition of animistic spirits — spirits in the trees and streams that had to be placated. Yet neither White's essay nor his writings on the history of technology present evidence that the doctrines of medieval Christianity led to the irresponsible use of nature. Environmentally insensitive opinions may certainly be found, as when, during German colonisation east of the Elbe, an abbot declared that 'the forest which adjoins Fellarich covers the land to no purpose, and hold this to be an unbearable harm'.[xxxviii] The arguments about belief systems remain inconclusive, since they omit to demonstrate that what the priesthood proclaimed, the peasantry carried into practice. The whole line of thought suffers from a lack of controls. To isolate the effects of Western Christianity, explicit contrasts would be needed with pre-Christian times and with behaviour in the non-Christian world.

With respect to pagan antiquity, the practical implications of spiritual thought are far from apparent. It was acceptable to fell a tree, mine a mountain or dam a stream provided one first made offerings to the spirit in charge. Sacred groves survive — islands in areas that have been deforested. Population growth was a greater influence than spirituality: it was real, whereas contrary to common opinion beliefs are continually being reshaped according to the exigencies of everyday life.[xxxix] What population growth did was oblige farmers to shorten the cycles of the shifting cultivation typical of the tropics, leaving ever less adequate intervals for the soil to recuperate. People whose values taught them the earth was sacred have often found that their urge to eat was a more compelling motive and have thereupon fudged their beliefs, contriving to leave their habitat a waste of erosion and deforestation.[xl]

The record of non-Western societies is largely ignored by writers on ecological history, apart from specialists on particular areas. Many ecologists continue to lay the blame for devastation on Western ideology. They tend to urge, hint, or take it as read that the entire Western economic system ought to be replaced before it does further harm. This is politics disguised as science. Yet other civilisations spread like stains across vast areas, with consequences for vegetation and wildlife that attract no censure. The West had its day when its share of world population was greatest; by that time its technology was the most powerful on earth. Yet these are matters of scale and timing, not fundamental differences in environmental practice. The transformations of global ecosystems are not being fairly assessed when they are presented as necessarily European, Western, capitalist or Christian.

ENDNOTES

[i] R. Lewinsohn, *Animals, Men and Myths* (New York: Harper & Bros., 1954), p. 126.

[ii] On the usage of penguins, see J. Sparks and T. Soper, *Penguins* (Newton Abbot: David and Charles, 1967), pp. 152–177.

[iii] H. R. Elder, *The Royal Fishery Companies of the Seventeenth Century* (Glasgow, 1912), p. 26.

[iv] J. Fisher and R. M. Lockley, *Sea Birds* (London: Collins New Naturalist, 1954), pp. 76–77; B. Heuvelmans, *On the Track of Unknown Animals* (London: Rupert Hart-Davis, 1958), p. 245.

[v] H. B. Cott, 'The exploitation of wild birds for their eggs,' *Ibis*, **95** (1953), pp. 443–444.

[vi] E. Lankester, *The Uses of Animals In Relation To the Industry of Man* (London: Robert Hardwicke [c. 1863]), p. 295.

[vii] F. G. Walton Smith and H. Chapin, *The Sun, the Sea and Tomorrow* (London: Hurst & Blackett, 1955), p. 46.

[viii] J. R. Baldwin, 'Sea Bird Fowling in Scotland and Faroe,' *Folk Life*, **12** (1974), pp. 61–62.

[ix] J. Fisher and R. M. Lockley, *Sea Birds*, p. 105. R. G. B. Brown, 'Fulmar Distribution: A Canadian Perspective,' *Ibis*, **112** (1970), pp. 44–51, argues that neither oceanographic factors (the distribution of macro-plankton) nor the distribution of fish offal account for the range of the Fulmar. This leaves the massive increase of population unexplained.

[x] Quoted by J. Yeats, *The Natural History of the Raw Materials of Commerce* (London: Cassell, Petter and Galpin, 1871), p. 10.

[xi] Quoted by W. Cronon, *Changes in the Land: Indians, Colonists, and the Ecology of New England* (New York: Hill and Wang, 1983), p. 109.

[xii] W. H. Pearsall, *Mountains and Moorland* (London: Collins, 1989), pp. 234–236.

[xiii] Bruhl Street, Leipzig, was accounted the absolute centre of world trade in furs, which is probably correct, even allowing for the Sino-Russian trade through Kiakhta.

[xiv] A. C. Laut, *The Fur Trade of America* (Norwood, Mass.: Norwood Press, 1927), p. 87; for doubts about other resource-based explanations of location, see E. Jones, *Locating the Industrial Revolution* (Singapore: World Scientific, 2010).

[xv] See E. Lankester, *Uses of Animals*; H. Poland, *Fur-bearing Animals in Nature and in Commerce* (London: Gurney and Jackson, 1892); E. M. Veale, *The English Fur Trade in the Later Middle Ages* (Oxford: Clarendon Press, 1966); J. Yeats, *Natural History of the Raw Materials.*

[xvi] E. L. Jones, *The European Miracle* (Cambridge: Cambridge University Press, 3rd edn., 2003), pp. 74, 81; D. Galton, *Survey of a Thousand Years of Beekeeping in Russia* (London: Bee Research Association, 1971), pp. 17, 61.

[xvii] K. Appuhn, 'Ecologies of Beef: Eighteenth-century epizootics and the environmental history of early modern Europe,' *Environmental History*, **15** (2010), pp. 268–287.

[xviii] M. Mancall, 'The Kiakhta Trade,' in C. D. Cowan (ed.), *The Economic Development of China and Japan* (London: George Allen and Unwin, 1964), p. 27.

[xix] E. Lankester, *Uses of Animals*, pp. 298–299. The English import of beaver furs was 80,000 per annum.

[xx] C. McEvedy and R. Jones, *Atlas of World Population History* (Harmondsworth, Middlesex: Penguin, 1978), p. 348.

[xxi] A. W. Crosby, *Ecological Imperialism: The Biological Expansion of Europe 900–1900* (Cambridge: Cambridge University Press, 1986).

[xxii] Contrary to the much-cited 'tragedy of the commons' publicised by Garrett Hardin; see Jones, *Locating*, pp. 218–219.

[xxiii] J. Rowe, *Cornwall in the Age of the Industrial Revolution* (Liverpool: Liverpool University Press, 1953), p. 272.

[xxiv] K. Appuhn, 'Ecologies of Beef', passim.

[xxv] C. David Badham, *Prose halientics* (London: John W. Parker & Sons, 1854), p. 445.

[xxvi] E. Lankester, *The Uses of Animals*, p. 231.

[xxvii] Anonymous, 'The Cultivation and Collecting of Medicinal Plants in England,' *British Medical Journal*, 31 October 1914, p. 760.

[xxviii] *Cf.* 'Wiping out wildlife "may not matter",' *The Independent*, 25 October 1997, referring to work by R. May and S. Nee.

[xxix] *New Scientist*, 15 March 2011.

[xxx] S. Briscoe and H. Aldersey-Williams, *Panicology* (London: Penguin, 2009), p. 214.

[xxxi] *New Scientist*, 21 May 2011.

[xxxii] *Financial Times*, 1 November 1997.

[xxxiii] P. M. Vitousek *et al.*, 'Human Domination of Earth's Ecosystems,' *Science*, **277** (1997), pp. 494–499.

[xxxiv] D. W. Steadman, 'Prehistoric extinctions of Pacific birds: Biodiversity meets zooarchaeology,' *Science*, **267** (1995), pp. 1123–1131. A 2013 study raises total extinctions of non-passerine land birds on the Pacific islands to nearly 1,000 but is based on extrapolation because the fossil record is so incomplete. (*Financial Times Magazine*, 30 March 2013).

[xxxv] J. C. Greenway, Jr., *Extinct and Vanishing Birds of the World* (New York: Dover Books, 1967), p. 7.

[xxxvi] *Ibid.*, p. 29.

[xxxvii] W. K. Hancock, *Professing History* (Sydney: Sydney University Press, 1976), pp. 151–152.

[xxxviii] Quoted by E. M. Veale, *English Fur Trade*, p. 172, n. 3.

[xxxix] E. L. Jones, *Cultures Merging: A Historical and Economic Critique of Culture* (Princeton: Princeton University Press, 2006).

[xl] J. D. Hughes, *Ecology in Ancient Civilizations* (Albuquerque: University of New Mexico Press, 1975), p. 150.

12. PRISTINE AMERICA

> The history of the recent extinction of birds on the North American Continent is so closely related in time to the penetration of the continent by Europeans and their civilization that it is impossible not to believe that the one is the result of the other.
>
> James C. Greenway, *Extinct and Vanishing Birds* (1967)

The Americas were already occupied when the colonists arrived. Although European occupation eventually raised exploitation to unprecedented levels, the New World was no *tabula rasa*. I shall illustrate the processes by means of two of the most significant movements that settlement entailed. They were the colonisation of New England and the surge onto the prairies.

The story opens with a brace of related myths, *passé* in scholarly circles but apparently ineradicable from the popular imagination. One is the Pristine Myth, which portrays the continent as crammed with unexploited wildlife and resources when the Europeans arrived. Ecstatic reports by the explorers help to convey this vision, but caution is required because they were keeping half an eye on impressing investors back home. Yet there is no warrant for dismissing their enthusiasm out of hand: the numbers of birds and animals and tall stands of timber rightly impressed the first comers.

Brave attempts have been made at estimating just what the scale of biological resources was. The vegetation is reckoned to have consisted of 913 million acres of forest (48% of the land surface), 723 million acres of grassland (38%), and 266 million acres of desert scrub (14%).[i] Of the forest, 150 million acres were destroyed by fires set by the settlers to open up cropland or pasture. By the 1940s, approximately 40% of the total land surface had been cleared for farming. The remaining 60% was mostly under some form of woodland cover, though much altered: only 100 million acres (11%) of the original forest remained. (Another author says that only 19 million acres or 4.4% of the original forest were left standing east of the Mississippi).[ii] Three-quarters of the land classified as forest lay in the eastern states and most of that was second growth. Surviving forests contained only one-third as much usable saw timber as had existed when the Europeans came and four-fifths of all the remaining saw timber lay west of the Great Plains.

When the Europeans landed, there were supposedly 60 million buffalo, 60 million beaver, 40 million white-tailed deer, 40 million antelope, 10 million elk, 2 million wolves and 500,000 black bears. Most of the buffalo were on the prairie but 5 million lived in the eastern woodlands, where the final herd was extinguished on the very last day of the eighteenth century.[iii] For good measure, there were said to be one billion squirrels, presumably an American billion, i.e., a mere 1,000 million! The fact that these figures embrace part of Canada (and for the antelope include Mexico), is scarcely here or there given the unavoidable degree of approximation. As late as 1900, there were 150 million waterfowl but by 1935 only 30 million remained. Before concluding that this rather implausible slump resulted from drainage or market-hunting, we should note that the 1930s were times of drought, which would have reduced the total. It is not worth cavilling about the figures: they can be only the roughest guides but they are guides nevertheless. If the true numbers differed substantially from them, they must still have been gargantuan.

Detailing the prodigious wildlife at the time of contact is the aim of an eloquent book, *Paradise Found*, by Steve Nicholls.[iv] I shall refrain from transcribing his material, which deserves to be read by anyone interested in ecological history. Descriptions of profusion drawn from original sources are reported species by species and region by region. Interactions among species are described, involving complex competition, unintended consequences, trophic cascades and so forth. Nicholls is as informative about marine and coastal changes as about terrestrial nature but I shall not dwell on those aspects.

THE PRISTINE MYTH

The reduction of wildlife through clearing the forest and hunting was on a vast scale. It was faster and far more market-driven than might be supposed. Virtually from the start, the Puritans exported resources.[v] 'Nor could it be imagined', wrote an observer in 1653, 'that this Wilderness should turn a mart for merchants in so short a space, Holland, France, Spain and Portugal coming hither for trade'.[vi] The immediate impact of the tiny communities of settlers was, in some ways, shocking. The hordes of

Passenger Pigeons inhabiting Massachusetts in the 1640s had been 'much diminished' by netting as early as 1672 and the wild turkeys had been wiped out too.[vii] In 1708, some New York counties felt they had to establish a close season for heath hen, grouse, quail and turkey, while two years later Massachusetts prohibited the use of boats to pursue water fowl.

The mix of species altered greatly once settlers broke up the forest, though by no means all changes were loss. Some birds and animals retreated fast but others were favoured by the increased ecotone (the edge between woods and cleared land). *The Skeptical Environmentalist* points out that forest clearance was a great natural experiment in modifying the environment. In the eastern United States, primary forest was reduced by approximately 98% or 99% over two centuries.[viii] This might have been expected to cause great and permanent losses, yet resulted in the actual extinction of only one forest bird. Paradoxically, if we assume this was the huge ivory-billed woodpecker, and accept contrary to all romantic hopes that it has finally gone, the reason for its loss may have been more complicated. Its habitat deteriorated with the end of the burning that the Indians had done, the forests becoming choked with scrub; admittedly loggers took out the remaining big trees that the ivory-bill needed for nesting.

An authority asserts that one hundred species of North American birds cannot live in climax forest and another makes the claim that for every land bird less abundant than when the Pilgrims landed, five or six had become more numerous by 1946.[ix] This catches the direction of change, though it may no longer be true to quite the same extent. One reason is that the winter habitat of summer visitors breeding in the United States and Canada has been fragmented by agricultural development in Central and Latin America. Another reason is over-browsing by white-tailed deer, which has destroyed the cover crucial for birds that nest on the ground or in the under-story in the eastern states.[x] As will be discussed in the final chapter, the emergence of a landscape of scattered woods and patches of corn has created more ecotone than ever, benefitting a burgeoning 'crop' of deer which is largely unharvested except as road kill. It brings nature to the very doorsteps of the suburban dwellers who live in a regrown forest.

The exact progress of forest reclamation will probably never be reconstructed, since biological records are few and not guaranteed to be representative. Settlers had preoccupations other than playing natural historian. For a long time, their society could not afford a class of full-time observers, though occasional men did record outstanding events, like passenger pigeons 'blackening the sun'. Treating white settlement as an undifferentiated episode of clearing dense forest, not to be reversed, would nevertheless be too dogmatic about the preceding uniformity of forest cover. Even so the change in cover was enormous. In the twentieth century, the Nobel laureate physicist, Murray Gell-Mann, and his brother used to return to the Bronx for bird-watching after their family had moved to Manhattan. Just north of Bronx Zoo was the one remaining stretch of the hemlock trees that had covered all New York. Displaying precocious ecological insight, Gell-Mann wrote that, 'Ben and I regarded the city as a hemlock forest that had been over-logged'.[xi]

THE NOBLE SAVAGE MYTH

The second myth is that of the Noble Savage. It claims that the indigenous inhabitants had scarcely imposed themselves on the bounty around them. A variant is that they were so experienced they were able to teach the clumsy Puritans how to live off the land. Either way the existing inhabitants are depicted as peaceable fellows who subsisted while insignificantly disturbing the animals. They ate but left no crumbs. On the contrary, Nicholls records the ways in which they had modified the environment.

Critics of the western world are eager to concentrate on the losses and urge that other cultures treated nature more kindly. The aggrieved tone was set by Lynn White's paper.[xii] Research, though, shows that the noble savage myth is just what it seems, a myth. Earlier and non-Western societies abused the environment, through poor forethought and brutal, extravagant methods. 'It is my contention,' wrote Daniel Guthrie, 'that primitive man was no better in his attitude toward his environment than we are today and that the concept of primitive man living in harmony with nature is a serious distortion of the facts'.[xiii]

Archaeological work has confirmed the scale of pre-European environmental insults. There was nothing idyllic about clearing forests by fire-stick and hard-scrabble farming.[xiv] Yet protestation seems in vain. Popular imagination, alarmed by the rapaciousness of industrial society, sides with Lynn White and finds it hard to let go of the notion that Christian society was uniquely destructive. This puts the Puritans in the line of fire — demoting them from the pantheon of folk heroes where earlier generations of patriotic Americans had placed them. In contrast the Indians are raised up as paragons of ecological sensitivity.

Misunderstanding is all too likely. What can be dismissed is categorical moralising about differences between the behaviour of Indians and of white settlers. While the technology and markets that the settlers had behind them in Europe meant their impact was ultimately far greater, it is a *non sequitur* to praise Indian life as mystically in tune with nature. Nicholls offers interesting counter proposals. The latent function of Indian rituals — he implies — may have been to ration the kill and avoid the tragedy of the commons. They may have been intended to re-establish 'some kind of balance' — better expressed as permitting hunted populations to recover — after phases of over-exploitation. Despite the communalism believed to be typical of the Indians' social arrangements, there were instances where they established individual rights in fisheries and marked out stretches of the shore in order to privatise worm grounds (the larvae were eaten).[xv]

Little wilderness rested unmodified during the pre-contact period. In the boldest interpretations, such as in the work of William Denevan, extensive parts of North America are described as already densely settled and heavily disturbed.[xvi] There were earthworks, roads, fields and villages. Other sources remark on how much the settlers gained by taking over land which was cleared at their entry and (by early standards) already 'developed'. The suggestion that the resultant gain was a generation's worth of labour is probably not worth pursuing because it can be only speculation. Yet there was a free lunch for the Europeans. In the Mississippi country, the earliest whites followed trails imprinted on the landscape by Indian travois — loads fastened between two shafts and pulled by dog team or horse.[xvii]

Surprisingly, the country as a whole may have become less rather than more humanized by 1750 than it had been in 1492. Indian landscapes had largely disappeared from the east, and the ravaging of the native population by disease helps explain why. Conceivably there was more 'primeval' forest as late as 1850 than there had been in 1650, though it was surely more fragmented and its species composition wider.[xviii] Forests had been modified, grasslands created and erosion caused. Denevan quotes a specialist on the history of fire to support the heterodox conclusion that truly virgin forest did not exist in the sixteenth and seventeenth centuries and was a retrospective fancy of the late-eighteenth and early-nineteenth century mind.[xix]

One hesitation about Denevan's conclusion on the Indian impact, and the comparable interpretation by Bernard Powell, might be that they rely on instances of dense indigenous settlement and heavy disturbance drawn from here and there across the Americas throughout long stretches of time.[xx] Granted the correctness of their specific examples, might it not still be that there was too small an Indian population with too limited a technology to have had country-wide effects? Another study, this time of bird life around the New Mexican pueblos, does however support the view that low-tech Indian society could exert considerable force.[xxi] In New Mexico disturbance, field irrigation, cultivation and the insects attracted to crops supported avian diversity; more precisely the effects came from uniform crop habitats attractive to opportunistic birds, the breakdown of habitat barriers, and the draw-card of the ecotone. Over 70 species were present, including duck, geese, hawks, eagles, American kestrels, mourning doves, horned larks, common ravens and jays.

Studies of seventeenth-century New England confirm the impression that there had previously been an Indian impact of considerable force. Thomson's book on *The Changing Face of New England* describes the forests as honeycombed with fields where the Indians had cultivated corn (maize) and left other marks of use and occupation.[xxii] Birds and animals homed in on the Indian clearings; the inhabitants' response was energetic bird scaring and animal trapping, just as there was around settlements in England. Small boys were employed since everywhere their labour had a low opportunity cost.

Accusations of unprecedented exploitation by the pioneers also sit uneasily with an emphasis on their feeble preparedness for life in the new environment. By contrast, the Indians are praised for forming a continuum with nature, operating smoothly and doing no harm. Romanticism suffuses conceptions of their use of land and wildlife. This is unhelpful when trying to assess their true impact — deeply unhelpful when it is protected by rhetoric designed to close off other interpretations.

Charles Mann goes out of his way to vaunt Native American technology relative to European.[xxiii] He complains that Europeans evaluate the Native Americans in terms of their own technology, and find it wanting. He urges that the Europeans prevailed in America because they were kicking at an open door, the native population having been drastically thinned by disease. This was definitely a consideration but the point might have been introduced without denigrating the European technological achievement. The virtue of European and American technology was that it was not static but went on improving. Leaving aside the later industrial achievement, consider what Lewis and Clark carried as field gear at the start of the nineteenth century. Native American technology did not develop in the same way. To quote a comment of mine on a review article in which Peter Munz took a similar view of the Maori, 'the quality of openness, rooted in the polycentric nature of Europe's society, gave Europeans options for growth and expansion seldom inherent in societies with more static hierarchies'.[xxiv]

No-one is in the least doubt that aboriginal peoples could be superlatively skilled at hunting and related skills; they practised all the time. Mark Twain wrote an admiring piece about the prowess and nicely honed equipment such as boomerangs of the Australian aborigines. Nor need Indian hunting of species where a few dominant males monopolised the females have done much harm, since there would have been a harvestable surplus of non-breeding males. Rather the issues are: were the Indians inherently more adept at forest living than the whites; did they teach the settlers to live off the land and how to farm; and did they skim off only surplus wildlife that would have perished before the next breeding season — and do all this without leaving a mark on vegetation, forest cover or landscape?

To be categorical about environmental modification by the Indians is tricky: they were not literate and later comers have erased many of their physical traces. Botkin sets out the categories of scholarly opinion on the matter; he discounts writers who assume the Indians had little effect, citing how important a material wood was to them, their craft skills, and the role in their economy of grassy openings where big game could graze.[xxv] Despite being low-density populations, settlers and Indians came into conflict because both coveted the forest openings cleared by the Indians for hunting and growing crops. They met awkwardly along the pathways that the Indians had formed. Usurping the resources created by their predecessors does the settlers no credit but does not mean they were worse woodsmen or farmers. Nor does it indicate — the exact opposite — that the Indians had floated through America without modifying its ecosystems.

A majority of secondary opinion has shifted to thinking the effects were considerable. Burning in the tall forests meant that at the time of contact they possessed no understory. They were silent by day and their birds were owls, ravens, eagles, passenger pigeons and Carolina parakeets.[xxvi] Elsewhere, burning the woodland meant that grass grew thickly in open patches and attracted the deer, elk and so forth whose numbers so astounded the incomers. The abundance of life in these park-like woods had a dark side: it attracted predators, among which wolves stood out because Europeans already feared them. But by the turn of the eighteenth century, the Indian population had fallen sharply, through war, disease and displacement. One unexpected possibility is that the plains bison, hitherto kept out by Indian hunting, were able to push into the eastern forests, only to succumb within a century to the pressure of white settlers and hunters. When Indian burning stopped, woodland began to grow back over the glades.[xxvii]

There is evidence that the Indians had not only spared fruit and nut trees when clearing fields, but even planted them. In Nicholls's words, 'far from being a pristine wilderness, the eastern broadleaf forests were, in effect, one vast Indian orchard'.[xxviii] Nicholls, though, does cite an authority who puts a less benign gloss on the Indian achievement. Deer, raccoons, bears and turkeys — unintended or only partly intended beneficiaries of Indian environmental manipulation — were competitors for the wild fruits and posed threats to Indian maize fields. The different proportions of mammal bones at archaeological sites are consistent with a deliberate effort to eliminate animal competitors. Reports by explorers and settlers tend to confirm this. There are two scenarios: in one, the abundance of game in early times was due to Indian land management; in the other, game flourished after the decimation of Indian populations. The scenarios are hard to reconcile, especially as the first whites found plenty of both game and Indians.

The 'historic overkill' thesis is at the heart of understanding environmental change, since any evaluation of the Indians' role and the fate of their societies sets the scene into which white settlers intruded. Sometimes — but not always — there is evidence of a heavy Indian impress, followed by the collapse or extrusion of their populations, and a recovery of hunted animals perhaps sizeable enough to explain awestruck reports by the first white arrivals. Historical documentation is not quite comprehensive enough to justify a confident conclusion, other than that there was no unique Indian-wildlife equilibrium. Nicholls wisely concludes that the relationships varied from place to place and time to time.[xxix]

CULTURAL EXCHANGES

The suggestion that, for lack of skill and hunting equipment, the earliest white settlers could not soon feed themselves seems unlikely.[xxx] In the presence of abundant game, how many hunters would have been needed to supply their tiny settlements and how long would it have taken for others to copy them? Conventional wisdom says the colonists needed to borrow techniques of transportation, agriculture, hunting and fishing from the Indians. This supposedly refers to colonies of the Scots, Irish, French, Dutch and

Germans, as well as the English. Some of these may have been sufficiently uninformed to need instruction, though it is not clear for how long and to what extent.[xxxi] If the colonists came from cities, as some did, the point may be understandable, although seventeenth-century cities were small, rustic and close to the land in every sense.

One practice at which the Native Americans allegedly surpassed the settlers was 'noodling', that is catching fish by hand. According to Bilger, the first Europeans to try this would have been clumsy in contrast to the amphibious excellence of the Indians.[xxxii] He imagines the settlers exposing their inexperience by their habit of baptising the practice with different culture names in different states. A diverse lexicon does not in the least indicate unfamiliarity with the practice, however; it acknowledges the dear transport and consequent localism of life in the past. England had its own array of local terms, some of which survive to this day, such as 'tickling' and 'guddling'. Undoubtedly, the first settlers had to master American river conditions and species of fish but this can have required only a short spell of learning-by-doing and would not have bulked large in the sweep of colonial history. Many already had the basic skill; noodling was an old European practice too.

The supposed tutoring of the whites by Indians was discussed by the late Lynn Ceci in her contribution to a book entitled *The Invented Indian*.[xxxiii] This pointed title was selected because a number of scholars, including Ceci, had been criticised for taking an unromantic view of Native American practices — criticised at every level down to what lawyers call mere vulgar abuse.[xxxiv] Ceci's particular topic was the legend so prevalent in the historiography of the United States as to have become folklore, that an Indian called Squanto taught the settlers how to manure their corn fields with fish. Yet Squanto had a remarkable personal history which may well have given him a chance to see English practice with fish fertiliser. If he indeed taught the practice to the settlers, he could have been only rendering unto Caesar what was already Caesar's. Perhaps English settlers from inland districts did lack experience with fish manure but they intermingled in the new American villages with others from coastal counties who could perfectly well have informed them.

Ceci tries to outflank her critics by claiming that fish do not after all make good fertiliser, but the deficiency would probably not have been significant to poor farmers with few alternative sources of manure. They would have been glad to apply any organic material to the fields. In Suffolk it was traditional to catch fish for that purpose.[xxxv] In 1489, fish fry were being caught with 'unreasonable nettes and engynes', half being fed to pigs and the other half, 'they put and ley it in grete pyttes in to the grounde', i.e., manure pits. The Puritan inhabitants of New England employed the same method, dunging their ground with the plentiful cod, shad and probably alewives.[xxxvi] In Old England, the practice long continued and in the early nineteenth century thousands of bushels of fish were still being taken from the Thames for manure.

The claim about Squanto's pivotal role derives from a single letter of 1621, whereas, according to Ceci, later observers found the Indians did not use fish or other fertilizer, relying instead on shifting cultivation. Yet there are other writers who claim that clearing the forest with stone axes was so hard no one would have willingly moved to another stretch of forest, a point that neglects the fire-stick. Agreed, since some tribes did build permanent villages, shifting around may have been infrequent. If so, this was perhaps because, contrary to Ceci, the Indians did use fish fertiliser to keep up soil fertility.

One of the most penetrating memoirs of pioneer life reports that in Vermont the Indians built weirs close to where they raised corn and other crops, 'because what they didn't eat they used for fertilizer. Fish of a certain size, or a certain amount of fish, went into every hill of corn.'[xxxvii] Are we to suppose that the practice had spread from the Plymouth colony to Indians in Vermont or had perhaps been learned from other whites? It seems unlikely: the report comes from a closely observed account of the way Indians exploited their environment which lacks any hint that their methods were second-hand. Rather than balancing an inverted pyramid of cultural exchange on the head of Squanto's pin, the fairest conclusion would be that settlers and Indians, both living close to the land and with low agricultural productivity, had independently grasped the opportunity to use fish fertiliser.

Beyond their purported superiority at farming, the impression is often given of what has been called 'the Indian-as-Mystic'.[xxxviii] Indians are credited with ecological insights not available to Europeans. Yet this view has been shown to be an artefact of modern Western ideology; the speech of Chief Seattle which embodies it has been revealed as a forgery, having been written by a freelance speechwriter for the American Baptist Convention![xxxix] 'I believe all stereotypic views are wrong', says Powell, when considering 'what only can be fairly called ecologically insensitive practices by various American Indians'. He cites buffalo being deliberately driven over cliffs and the hunters wasting one in four of those killed. In efforts at bird-scaring, the Mohawk Iroquois suspended young crows alive. Far from being 'brother to the beaver', once the Indians acquired European tools and weapons, they pursued beavers to extinction in order to supply the market for furs.

Consider, too, the take-over of an Indian fish trap in Missouri: this trap was apparently constructed in the early nineteenth century, bought and improved by a settler in 1837 and kept in use until 1910.[xl] The trap must have been effective because that year butchers in nearby towns complained about competition from fish-mongering and, since trapping fish was illegal, armed game wardens went to destroy the trap. They dynamited it. But the trap was rebuilt and not finally destroyed until 1923, while another trap was not destroyed until the 1950s. In an imbroglio like this, attaching moral blame to one society rather than the other makes no sense. Both were trying hard to live off the land.

EUROPEAN SETTLEMENT

Europeans began at once to modify America's environment: they dumped flints brought as ballast overboard in the harbours. A trivial example, maybe, but suggestive, and in other respects the effects were massive, some of them seriously negative. The simplest interpretation of the impact of the settlers is that they vastly expanded changes of the types wrought by their predecessors. They enlarged Indian clearings and adopted the habit of moving on when crop yields began to fall or if firewood for the freezing New England winters could no longer be gathered nearby. A knee-jerk response that shifting cultivation was wasteful, repeated by European travellers in the early United States and by the ecologically minded today, takes no account of what was the scarce factor. For the settlers it was labour, not land, and they were not senseless in moving on, which they could do as long as the Indians could be pushed back. 'The idea of exhausting the soil by cropping, so as to render manure necessary, has not yet entered into the estimates of the western cultivator', Morris Birkbeck wrote in 1818, 'manure has been often known to accumulate until the farmers have removed their yards and buildings out of the way of the nuisance. They have no notion of making a return to the land; and as yet there seem no bounds to its fertility'.[xli]

As the colonial population grew, the prospects for shifting cultivation from plot to plot or place to place in the East became restricted.

Under repeated cropping, organic material was leached out of the soil. The desertion of New England farms and their replacement from the late nineteenth century by second-growth woodland of low productivity has been attributed to impoverished soil. But abandonment was so widespread, it could more reasonably be attributed to competition from the prairie states, which undercut New England's crops in the market. This was a nineteenth-century phenomenon; previously it had been common for entrepreneurs to clear land and supply it with the rudiments of capital stock, to let or sell to later comers. A characteristic advertisement placed in a colonial newspaper in 1775 read, 'Jonathan Moulton of Hampton in the colony of New Hampshire offers to let for a number of years twelve farms in the township of Moultonborough made within a few years from new and uncultivated lands...'[xlii]

By Moulton's day, indeed soon after the Pilgrims' arrival, aspects of American ecology began to be altered. The arrival of new crops and, along with them, weeds is well documented.[xliii] Uniform stands of crops and herds of cattle and pigs were magnets for pests. The colonists were obliged to go in for bird scaring and other means of protecting plants and animals, as the Indians had been and as they had been accustomed to do themselves in Europe. Once crops and livestock were at risk, upturns in the populations of harmful species meant efforts had to be made to cull them, an activity formalised in communal ring shoots that could be dressed up as recreation. The significance of these events is rarely noticed by economic historians, though in one recent source there is a casual mention of the 'great Kentucky squirrel hunt' of 1820, when an 'Emigration' of squirrels from across the Ohio River was destroying the corn fields.[xliv]

Before moving this account on, as it were, to the West, it is worth remarking some of the consequences of altered land use along the eastern seaboard. Perhaps typical were the results of settlement in North Carolina, where large predators decreased but smaller species, like raccoons, opossums, squirrels, rabbits and quail, increased.[xlv] Richard H. Pough charts the range extensions and contractions of birds, particularly noting species that have become commoner than they were at the time of the great bird painter, John James Audubon. Pough's comments about changes in habitat and distribution would be refreshing additions to any identification guide.[xlvi] By his account, range extensions consequent on the opening of the forest, new food sources, and the provision of nesting sites on man-made structures, outnumbered contractions resulting from ploughing grassland and draining swamps.

A major feature was the abandonment of farmland in the northeastern states, leaving large areas to sprout second-growth timber, with the emergence further south of mosaics of regrowth and farmland. Several species colonise the early stages of forest regeneration but do not persist once woodland becomes dense again. Regeneration was a second-order consequence of the western movement, occurring when the newly opened farmlands were linked to eastern markets, notably by the Erie Canal, and the original farms given up. Despite their closer proximity to the market, eastern farmers could not compete in growing grain and eventually many of them could not compete, period. They left and their land returned to timber.

Plainly, American environmental history is not solely a narrative of westward movement. Innumerable variations occurred, including two strong counterpoints, the kickback of organisms attracted by stands of farm crops and the reversion to forest once the cereal frontier had moved on. In New England, especially northern New England, woodland has reasserted itself over wide expanses. Stone walls through the woods mark what now seem aimless boundaries and every so often the holes of old farmhouse cellars appear. The characteristic natural history of a former farm in Connecticut, intended as a nature reserve and now open to the public, was described in detail by Edwin Way Teale.[xlvii] The process continued, for instance in Upstate New York, where in 1966, Edmund Wilson noted of a poor agricultural district that, 'in proportion as the human population here is giving up and abandoning the countryside, the wild beasts seem to be moving back in'.[xlviii] Not only the northeast was affected; all woodland in North Carolina, except on the steepest slopes and in the swamps, lies on abandoned farmland. Supposedly worn-out land was not necessarily deserted; small farms in the South were often thrown together as cotton plantations.

Elsewhere, with little fresh land available before the break-out to the West in the 1780s, there was an incentive to raise yields on the existing cultivated area. A dense and still fairly static American farming population began to behave like Europeans. In this connection, an interesting counter-factual statement appeared in an early nineteenth-century *Farmer's Register*: 'if there had been no western country... Virginia would already have reached a high state of agricultural improvement'.[xlix] Colonists were attracted to rotations being developed in England. There is evidence of agricultural intensification in Virginia, the Middle Colonies and Massachusetts during the second half of the eighteenth century, before much new land became available, or perhaps more precisely, before it entered into competition with existing farms.[1] Interest continued during the nineteenth century. Proximity to expanding markets, including east coast ports where immigrants poured in and whence food could be exported, did offer some compensation. But in the end, the supplies of cheaper food from the interior were overwhelming.

REACHING FOR THE WEST

Some years ago, there was a debate among historians over whether, or how far, colonial farming was commercialised. Generalising is difficult when dealing with thousands of units standing at different distances from markets and exhibiting every form of family and labour arrangement, as well as diverse ecological settings. But the limiting case of pure subsistence was rare, if it ever existed. Settlers wished to consume the goods that were customary in the countries from which they came and entered the market economy when they could. Semi-subsistence production decreased over time, first in regions with good access to towns, cities and ports. At the opposite extreme from subsistence, a few rare farms grew only two or three products and bought in everything the family consumed, as well as all necessary inputs.

Whatever the precise extent of commercialisation in the coastal colonies, the cost of transportation limited the role of the market. Frontier producers were disadvantaged. The Whisky Rebellion reflected the difficulty they faced in finding anything they could profitably carry for sale: they could not afford to haul loads of cereals overland but barrels of spirits could be packed on horses. Ginseng roots for the China trade were an even lighter product; Daniel Boone hunted for them. Yet as the frontier moved westwards, difficulties mounted. Across the Appalachians isolation threatened. An incipient Balkanisation of the young United States was signalled by the fleeting rise in 1784 of a would-be state known as Franklinia — a breakaway that was soon annexed by North Carolina.

The real reason Balkanisation was thwarted was because improved communications succeeded in keeping the trans-Appalachian lands attached to eastern and Atlantic markets. The markets were centred in London but included Iberia, whose merchants had very early come to colonial America to buy bread grain. This spurred the middle colonies to specialise in grain production, just as English industrial demand encouraged cotton growing in the Old South. The eastern seaboard is a long one, with a north–south sequence of climatic zones. Export crops evolved regional economies and had specific repercussions on local ecosystems, arrangements that might be consistent with staple theory. Different crop specialisations created a chain of commodity landscapes in which distinctive associations of plants, weeds, animals and birds were bred. Early natural history records are sparse but the diversity can be spotted in regional literatures including *Walden* and *To Kill a Mocking Bird.*

The Westward movement was a feat of social organisation with the success of early land sales only outclassed by the cheap disposal of railroad land that brought a torrent of hopeful farmers into the interior. From the start of the nineteenth century, surveys of the wilderness were followed by 40 years of mass auctions, giving rise to the expression 'a land office business.' The system of surveying, hearing claims, and arranging or re-arranging occupation of the land was established by Albert Gallatin, and was reminiscent of the enclosure movement in England — except this time it was designed to allow the small man his own farm rather than divest him of it.[li]

Land in small parcels was sold by the surveyor at the Holland Land Company, Batavia, New York, against the wishes of the absentee owners who were investors from Amsterdam. They had bought an extensive area of western New York and would have preferred to sell in large lots, as did happen along the Hudson Valley. The history of the Hudson Valley damages the myth of America's universally free or readily purchasable land, since it fell into the hands of patroons like Van Rensselaer and perpetuated a quasi-feudalism long since given up in Holland itself.[lii] The leases offered in the patroonships reserved one-quarter of each plot to the landlord and tied the occupants to the soil. This was not true feudalism because that would have involved mutual obligations conspicuously absent in New York State, but it was sufficiently unjust to provoke a violent anti-rent movement, with gun-fights in the 1860s.

In the broader history of the Western movement, the patroonships were the exception to prove the rule, a regressive European planting in the New World. The usual American arrangement was the rapid sale of lots, following a rectangular land survey that imposed a grid of fields on the landscape and grids of streets on the towns.[liii] For anything comparable, Europeans would have had to hark back to Roman times or at any rate to the bastides, planned towns of the High Middle Ages. American settlers were able to establish farms quickly, wherever possible choosing sites with enough timber for woodlots, and proximity to roads, rivers and later railroads for shipping out their produce.

As the availability of cheap, fertile glacial till became better known, population built up in the Old Northwest, south of the Great Lakes. The United States came to possess a commercial agriculture with all the ecological effects that follow from concentrating on a few crops. The collecting and redistributing centre was Chicago — *Nature's Metropolis* is the title of William Cronon's blockbuster about it.[liv] From Chicago unimaginable vistas of tall grass prairie stretched away, while the north woods were within reach. Forty years later, it was said, 'we cannot but imagine the valley of the Mississippi as a huge farm with a very small grove in the northeast corner'.[lv] Very small relative only to the size of the valley! Chicago might be best known for amassing the wheat and corn grown on the former prairie, and processing meat from the cattle that replaced the herds of buffalo, but for a time the city was also the lumber capital of the world. Never, it seemed, could these resources run into diminishing returns.

This was the workaday, stockyard city that Carl Sandburg called 'Hog butcher for the world'. The boast was proud. A telling story is one repeated by Cronon about the New Zealander converted to understanding that the geysers of Rotorua were no achievement compared with watching a man stand up to his waist in blood sticking pigs in the stockyards. Chicago was the lynchpin of huge ecological zones, radically simplified in order to be made fruitful, mainly by putting the native grassland under crops. Natural resources, untended, did eventually run out. Unlikely though it had seemed, even the teeming herds of buffalo were eliminated in favour of almost as many cattle. The farms remained, the prairie grass replaced by sown pasture or fields of grain.

Americans, wrote a perspicacious lumberman in his 1915 autobiography, assumed they had made themselves great, 'whereas their greatness has been in large measure thrust upon them by a bountiful providence.'[lvi] The gibe has been made many times since, but is it correct? Cheap resources were an enormous boon that pushed the cost of material inputs lower than Europeans could conceive and lower than even East Coast Americans could hope for — yet what is to say the resources would be used? They were an inducement to exploitation, but every society must fashion its response. The Indians used them relatively lightly, European Americans used them heavily, and few took heed for the morrow.

American ingenuity lay in organising exploitation, for instance, by establishing the Chicago futures market in grain as early as the 1860s. When one resource was burned out they replaced it with some fresh enterprise. In American history, resources have been fungible, that is to say one has been substituted for another. Market power overrode ecological circumstances. 'Species thrived more by price than by direct ecological adaptation'.[lvii] Chicago remained the price-setting mart for raw materials. By hysteresis, the city outlived much of the bounty of nature which had given it so profitable a start. It still consumed immense resources but was capable of profiting from the generalizable skills developed by marketing, storage and redistribution.

REACHING THE PRAIRIE EDGE

'The view of that noble expanse was like the opening of bright day upon the gloom of night', was the declaration in 1818 by Morris Birkbeck, one of the founders of the settlements on the Wabash.[lviii] He came from Wanborough, Surrey, and by the time he reached the prairie, his feeling was of relief after having been 'long buried in deep forests'. Unlike Americans, the English did not have two recent generations of experience clearing timber and made heavy weather of the new task. Neither people had yet faced the opposite problem, a total absence of trees. The new challenge arrived in the region of Chicago. The first settlers described the woods ending as if reaching a wall, although modern scholars insist they degenerated into isolated groves and thereafter into sloughs and river bottoms. Either way, then came the prairie.

The pioneer and his wife felt the transition as a geological unconformity, a perverse one, because they had to have lumber for cabins, fences, and to feed the fire. One much-married man in Illinois promised likely women that, 'they should live in the timber where they could pick up their own firewood'.[lix] Such a man might halt at the forest margin, but others pressed on, casting about for wood lots. Once out on the prairie, it was necessary to keep a fire in at all times or venture to the next house for coals — not an enticing prospect in a Middle Western winter. One family in Benton County, Indiana, arrived from Pennsylvania via a district in southern Indiana which a descendant described to me as worthless by the twentieth century, but possessed of the nineteenth century essentials: wood and water.[lx]

In prairie circumstances, a grove was a prime site, offering timber yet abutting onto grassland that could be cultivated.[lxi] Further out, with more people and fewer trees, it became clear that to secure the timber for settling the prairie it would be necessary to raze the north woods, carry the lumber down to Chicago, and redistribute it from there. A rare surviving example of a grove is Kennicott's Grove, Glenview, Illinois, now encased in Chicago's outer suburbs.[lxii] It was settled by the father of the first curator of the natural history museum at Northwestern University, Robert Kennicott, whose appraisals were what persuaded the United States to buy Russian America and incorporate it as the state of Alaska. The Grove now houses a museum of natural history and an educational centre, in a tract of open woodland where (unusually this far west) scientific observations stretch back more than 150 years. Not far away, also in suburban Glenview, is Peacock Prairie, one of the last fragments of prairie in the Chicago region. Of the 21 million acres of natural grassland original to Illinois, only bits and pieces survive: here is a precious five acres.[lxiii]

On the sole occasion that Peacock Prairie was farmed, by a little light grazing in 1926–1927, a few invasive weeds were found. By 1967 they were closed out.[lxiv] Native vegetation is resilient and introduced plants are dependent on continued disturbance. Spectacular though the spread through America of many plants, animals and birds has been, it is the alteration of habitats to make niches for them that is the key to their success.[lxv] Ploughing the bulk of the prairie gave alien species their opportunity and began something like successive ecological detonations. Nevertheless, the current retreat of farmland in far parts of the Western plains may reverse the process and establish retro-landscapes.

No more lyrical natural history of the prairie is to be found than John Madson, *Where the Sky Began.* It is full of detailed ecology, lightened by personal touches and populated by real people. The chapter, 'A Prairie Bestiary', considers gains and losses from breaking-up the tall grass and indicates the species likely to have been present when this started about 1830. The waves of change come out clearly, some species shrinking away and being replaced, or expanding at first only to shrink later. Rise and fall was the fate of the prairie chicken, a grassland native that swelled to huge flocks on discovering corn and wheat as food in the late nineteenth century, then subsided when the land went so much under cereals there was just too little grass left.

The ranking historian of the High Plains is James C. Malin, who worked in the mid-twentieth century.[lxvi] Malin is claimed today as a pioneer historian of the American West and its environment, but if imitation is the sincerest form of flattery he is not much flattered: his approach is out of joint with environmentalism. The view gained currency in his day, at first only tacitly, that aboriginal man was in equilibrium with nature. So-called civilised man was regarded as destructive and out of balance.

Malin pinpointed these opinions as conceptual barriers. They got in the way of candid interpretation. Indian hunting of the bison never stabilised. There was no single Indian-wildlife equilibrium, although there may have been multiple equilibria, each sensitive to the possibility of further disturbance.

Malin thought the view that mesquite, sagebrush and cactus had spread through over-grazing was wrong or exaggerated; they were already present throughout their modern range in the period 1820 to 1850, before any major impact by cattle. He also thought the Plains Indians had entered the region too recently and too abruptly to have established a stable equilibrium with the environment. Before white settlement, Indian culture had already been derailed by adopting the horse and by a fluctuating distribution of tribes. Malin believed the tall grass plains were subject to long-term variation at the hands of natural factors, such as fires caused by lightning strikes. He documented severe dust storms in years before ploughing the grassland could possibly be blamed. Imitating Indian ways would not, he thought, be safe. Indian culture may have been heading towards an environmental crisis of its own before it was suppressed by the white man and Malin had the temerity to suggest that displacement provided the Indian with a good alibi. He was unimpressed by the environmentalists' wish to restore pre-contact ecosystems that he suspected were ideological constructs rather than honest pictures of the High Plains.

The Great Plains occupy one-fifth of the area of the lower 48 states. The plough reached the most distant Plains in eastern Montana about 1909 but today a dozen farms have been abandoned for every one still occupied.[lxvii] Most have collapsing buildings on them. In Nebraska, 5,000 or 10,000 farm houses stand abandoned, 25,000 in South Dakota, a majority still with structures.[lxviii] Responding to the challenge, in 1987, Frank and Deborah Popper (planner and geographer from the east coast) proposed that agricultural decline should be carried to the conclusion of deliberate retreat.[lxix] They showed that 109 counties contained a reducing population of only 400,000 out of a Plains states' total of 6.5 million. The farmers were defeated by recurrent drought and the Poppers were convinced that water shortage would ultimately heap disaster on those remaining.

Their solution was as bold as it was unpopular. One writer called the Popper's idea the biggest step in redefining America since the Alaska Purchase. Abandon, they said, a vast north–south swathe of the dry country and make it a massive ecological reserve, where the buffalo roam! They argued that in 1895 predicting the forests of New England would return as second-growth would have seemed deranged, yet it came to pass: wanting the buffalo back on the Plains is no more fanciful. Their proposal was denounced for 20 years and has faded a little from view because there are countervailing movements based on finding oil and on high commodity prices.

Families who quit their Montana farms between 1917 and 1922 were following the custom of the country: when the forest is logged or the seam mined to the point of diminishing returns, you move on. 'When, eventually, you reach the coast, you go trawling for salmon in Alaska'.[lxx] But even the stock of Alaskan salmon could run out. Jonathan Hughes, who worked in Alaska as a fisheries accountant, observed that, 'the old idea that salmon were naturally a renewable resource and a community property had been a blue-print for abuse and ultimate exhaustion. But in 1951 that was not yet clear'.[lxxi] It soon was, in this and many other respects.

ENDNOTES

[i] W. R. Van Dersal, *The American Land: Its History and Its Uses* (Oxford: Oxford University Press, 1943), pp. 19, 158, 159; citing on land use, H. L. Shantz and R. Zorn in the *Atlas of American Agriculture*; and on big game, E. T. Seton, *Lives of Game Animals.* An excellent survey of wildlife history by habitat type and species is P. Matthiessen, *Wildlife in America* (New York: The Viking Press, 1964).

[ii] J. C. Greenway, *Extinct and Vanishing Birds of the World* (New York: Dover, 1967), p. 37.

[iii] Evocatively described by B. Gilbert, *In God's Countries* (Lincoln: University of Nebraska Press, 1984), pp. 77–90.

[iv] S. Nicholls, *Paradise Found: Nature in America at the Time of Discovery* (Chicago: University of Chicago Press, 2009). Works on North American environmental history, once largely ignored by mainstream scholars and still far from integrated into general history, are now appearing in rapid succession.

[v] The environmental literature scarcely encourages discussion of Spanish and other early incursions and I shall concentrate, conventionally enough, on Puritan and other northern European activities, while recognising the force of the arguments in T. Horwitz, *A Voyage Long and Strange* (New York: Picador, 2008).

[vi] Quoted in W. Cronon, *Changes in the Land: Indians, Colonists, and the Ecology of New England* (New York: Hill and Wang, 1983), p. 167.

[vii] P. Mathiessen, *Wildlife*, pp. 56, 64–66, 281.

[viii] B. Lomborg, *The Skeptical Environmentalist* (Cambridge: Cambridge University Press, 2001), p. 254.

[ix] R. H. Pough, *Audubon Land Bird Guide* (Garden City, New York: Doubleday, 1946), p. xxvii.

[x] L. Line, 'Silence of the songbirds,' *National Geographic*, **183** (6) (1993), pp. 68–91.

[xi] M. Gell-Mann, in J. Brockman (ed.), *When We Were Kids: How a Child Becomes a Scientist* (London: Jonathan Cape, 2004), p. 36.

[xii] L. White, Jr., 'The historical roots of our ecologic crisis,' *Science*, **155** (1967), pp. 1203–1207.

[xiii] D. A. Guthrie, 'Primitive man's relationship to nature,' *Bioscience*, **21** (1971), p. 721.

[xiv] A large literature now exists to show how much forest clearance was undertaken by small-scale early societies, with dire results in soil erosion and the extermination of the larger species of wildlife. See, for example, *New Scientist*, 24 February 1990; J. D. Wilde, 'Archaeology and mass extinctions in New Zealand,' *1992 Yearbook of Science and the Future* (Chicago: Encyclopaedia Britannica, 1991), pp. 282–283.

[xv] S. Nicholls, *Paradise Found*, p. 377.

[xvi] W. M. Denevan, 'The pristine myth: The landscape of the Americas in 1492,' *Annals of the Association of American Geographers*, **82** (3) (1992), pp. 369–385.

[xvii] H. Binder Johnson, *Order Upon the Land: The U.S. Rectangular Land Survey and the Upper Mississippi Country* (New York: Oxford University Press, 1976), p. 83.

[xviii] In Wisconsin, the total woodland perimeter, so attractive to many bird species, rose by 240% between 1882 and 1950. See W. L. Thomas (ed.), *Man's Role in Changing the Face of the Earth* (Chicago: University of Chicago Press, 1956), II, p. 728.

[xix] W. M. Denevan, 'Pristine myth,' p. 380, quoting S. D. Pyne, *Fire in America* (1982).

[xx] W. M. Denevan, 'Pristine myth'; B. W. Powell, 'Were these America's first ecologists?' *Journal of the West*, **XXVI** (1987), pp. 17–25.

[xxi] S. D. Emslie, 'Birds and prehistoric agriculture: The new Mexican pueblos,' *Human Ecology*, **9** (1981), pp. 305–329.

[xxii] B. Flanders Thomson, *The Changing Face of New England* (Boston: Houghton Mifflin, 1977), pp. 19–21.

[xxiii] C. C. Mann, 'Native ingenuity,' *Boston Globe*, 4 September 2005.

[xxiv] E. Jones, *The Record of Global Economic Development* (Cheltenham: Edward Elgar, 2002), p. xi; P. Munz, 'The two worlds of Anne Salmond in postmodern fancy-dress,' *The New Zealand Journal of History*, **28** (1) (1994), pp. 61–62.

[xxv] D. B. Botkin, *Our Natural History: The Lessons of Lewis and Clark* (New York: Perigee Books, 1996), p. 236.

[xxvi] E. L. Jones, 'Creative disruptions in American agriculture, 1620–1820,' *Agricultural History*, **XLVIII** (4) (1974), pp. 514–515 and n. 16 for sources.

[xxvii] See example, D. Connery, *One American Town* (New York: Simon and Schuster, 1972), pp. 26–27, writing about Connecticut.

[xxviii] S. Nicholls, *Paradise Found*, p. 122.

[xxix] S. Nicholls, *Paradise Found*, p. 130 and passim.

[xxx] E. A. Bergstrom, 'English game laws and colonial food shortages,' *New England Quarterly*, **XII** (1939), pp. 681–690.

[xxxi] M. Morris, in a review in *Journal of World History*, **22** (2) (2011), p. 397.

[xxxii] B. Bilger, *Noodling for Flatheads* (London: William Heinemann, 2000), p. 17.

[xxxiii] L. Ceci, 'Squanto and the Pilgrims: On Planting Corn "in the manner of the Indians",' in J. A. Clifton (ed.), *The Invented Indian* (New Brunswick, NJ, USA: Transaction Publishers, 1990), pp. 71–89.

[xxxiv] See the Appendix, 'Criticisms of Nonconformers,' in Clifton (ed.), *Invented Indian*.

[xxxv] Anonymous, *The Fish Trades Gazette*, 19 March 1921, p. 27.

[xxxvi] T. Morton, *New English Canaan* (1632), quoted in W. H. Rowe, *The Maritime History of Maine* (New York: W. W. Norton, 1948), p. 20.

[xxxvii] W. Needham and B. Mussey, *A Book of Country Things* (Lexington, MA, USA: The Stephen Greene Press, 1965), p. 141.

[xxxviii] B. W. Powell, 'Were These America's First Ecologists?' *Journal of the West*, **XXVI** (1987), pp. 17–25.

[xxxix] In B. De Vries and J. Goudsblom (eds.), *Mappae Mundi* (Amsterdam: Amsterdam University Press, 2002), p. 367.

[xl] W. F. Shields, 'The Charlton River Fish Trap,' *Missouri Historical Review*, **LXI** (1967), pp. 489–496.

[xli] M. Birkbeck, *Letters from the Illinois Territory* (Belfast: F. D. Finlay, 1818), p. 29.

[xlii] *The New-England Chronicle* (or) *the Essex Gazette*, 28 December 1775 to 4 January 1776. Buildings already erected were listed. The entrepreneur was Col. Jonathan Moulton, an energetic breaker-in of fresh land.

[xliii] E. L. Jones, 'Creative disruptions.'

[xliv] In J. Horn *et al.* (eds.), *Reconceptualizing the Industrial Revolution* (Cambridge, MA, USA: The MIT Press, 2010), p. 210.

[xlv] S. A. Marks, *Southern Hunting in Black and White: Nature, History, and Ritual in a Carolina Community* (Princeton: Princeton University Press, 1991), p. 33.

[xlvi] Pough, *Audubon Land Bird Guide*, passim.

[xlvii] E. Way Teale, *A Naturalist Buys an Old Farm* (New York: Ballantine Books, 1975).

[xlviii] E. Wilson, *Upstate: Records and Recollections of Northern New York* (London: Macmillan, 1972), p. 237. For a first-hand account of the wildlife effects in Connecticut, see C. Rand, *The Changing Landscape: Salisbury, Connecticut* (New York: Oxford University Press, 1968), pp. 56–75.

[xlix] Quoted by A. O. Craven, 'Soil Exhaustion as a Factor in the Agricultural History of Virginia and Maryland, 1606–1860,' *University of Illinois Studies in the Social Sciences*, **13** (1925), p. 124.

[l] *Cf.* E. L. Jones, *Agriculture and Economic Growth in England 1650–1815* (London: Methuen, 1967), p. 48; H. L. Kerr, 'Introduction of forage plants into Ante-Bellum United States,' *Agricultural History*, **XXXVIII** (1964), pp. 87–95.

[li] M. J. Rohrbough, *The Land Office Business: The Settlement and Administration of American Public Lands, 1789–1837* (New York: Oxford University Press, 1968), especially p. 39.

[lii] H. Christman, *Tin Horns and Calico: An Episode in the Emergence of American Democracy* (New York: Collier Books, 1961).

[liii] B. Johnson, *Order Upon the Land.*

[liv] W. Cronon, *Nature's Metropolis* (New York: W. W. Norton, 1991).

[lv] Quoted in W. Cronon, *Nature's Metropolis,* p. 154.

[lvi] Quoted in W. Cronon, *Nature's Metropolis,* p. 206.

[lvii] W. Cronon, *Nature's Metropolis*, p. 266.

[lviii] Quoted in B. Johnson, *Order Upon the Land,* p. 17. See also M. M. Quaife (ed.), *Pictures of Illinois One Hundred Years Ago* (Chicago: Donnelly & Sons, 1918).

[lix] W. Havighurst (ed.), *Land of the Long Horizon* (New York: Coward-McCann Inc., 1960), p. 194.

[lx] Information from R. A. Swan, abstractor, Fowler, Indiana, and documents in his possession when visited on 22 January 1966.

[lxi] The complexity of ecology, settler origins, and date, is emphasised by T. G. Jordan, 'Between the forest and the prairie,' *Agricultural History*, **38** (4) (1964), pp. 205–216. The distinction between dense forest and open prairie was blurred by a belt of mixed ecosystems.

[lxii] I claim credit for the preservation of The Grove in the 1970s, together with two campaigners known condescendingly to the would-be developers as the 'Frog and Fern ladies' (Donna Ventura and Dee Stowe). See E. Jones, 'Saving Kennicott's grove: Successful conservation at an American Coate,' *The Richard Jefferies Society Journal*, **23** (2012), pp. 6–16.

[lxiii] Compare Botkin's search for the fragment which is now the Alwine Prairie Preserve of the University of Nebraska, Omaha. Botkin, *Our Natural*

History, pp. 258–259. J. Madson, *Where the Sky Began: Land of the Tallgrass Prairie* (Boston: Houghton Mifflin, 1982), pp. 287–302, has an extensive appendix of prairie preserves as at 1979.

[lxiv] E. L. Jones, 'Creative disruptions,' p. 513, n.8.

[lxv] A study of fish in California, where 50 species have been introduced since 1871, concluded that declines of native fish were largely due to habitat change, to which admittedly the introductions may have contributed. P. B. Moyle, 'Fish introductions in California: History and impact on native fishes,' *Biological Conservation*, **9** (1976), pp. 101–118.

[lxvi] J. C. Malin, *The Grassland of North America: Prolegomena to its History With Addenda and Postscript* (Gloucester, MA, USA: Peter Smith, 1967 [1st edition, 1947]; J. C. Malin, 'The Grassland of North America: Its Occupance and the Challenge of Continuous Reappraisals,' in W. L. Thomas (ed.), *Man's Role in Changing the Face of the Earth* (Chicago: University of Chicago Press, 1956), pp. 350–366.

[lxvii] J. Raban, *Bad Land* (London: Picador, 1997).

[lxviii] P. L. Brown, 'Ghost houses reflect fading of farm life,' *New York Times*, 2 April 1992.

[lxix] A. Matthews, *Where the Buffalo Roam: The Storm Over the Revolutionary Plan to Restore America's Great Plains* (New York: Grove Press, 1992).

[lxx] J. Raban, *Bad Land*, p. 229.

[lxxi] J. R. T. Hughes, 'The Great Strike at Nushagak Station, 1951: Institutional gridlock,' *Journal of Economic History*, **42** (1982), p. 5.

THE MODERN WORLD

13. EAST ASIA

> Were it not for human impact Hong Kong's forests might still support elephants, tigers, rhinoceroses, gibbons, pheasants and woodpeckers.
>
> *Biodiversity Survey of Hong Kong* (2002)

CHINA'S ENVIRONMENTAL PAST

The treatment of natural environments in Asia has been, and remains, of signal importance to the environmental history of the world. The subject has often been misunderstood, not least because Western understanding of Asian attitudes was for a long time suffused with romanticism. This can be traced back to Montesquieu, if not before. A vision of an orderly society ruled by a benign emperor was appealing amidst the contentiousness of eighteenth-century Europe, where religious history had been savage and bigotry persisted.[i] Romanticism reappeared in the late twentieth century, when it was based on the delusion that Asians knew how to live in harmony with nature: this was an inverted projection of Europe's own attitudes, not unlike the Noble Savage myth foisted on Native Americans. Views like these are fantastical but hard to pin down, since beliefs, and *a fortiori* behaviour, have seldom been stable from one period to the next.[ii] Even Buddhist writings are ambivalent about the sanctity of the natural world, while the downside of the way in which Shintoism grants wildlife equal status with people is that, like humans, animals are expected to look after themselves.[iii]

China, much the largest country in East Asia and the fount of much of the region's culture, never experienced the environmental peace of legend. It was stripped of forest, provoking run-off, erosion and a loss of nitrogen from uplands colonised by land-hungry peasants. It took a Chinese scholar, Yi-fu Tuan, to contradict the uncritical praise of Taoist and Buddhist traditions by demonstrating that they were not matched by practice.[iv] Memorials in favour of conserving resources were presented to emperor after emperor but this merely indicates the frequency with which they were disregarded. Chinese emperors could be as keen as any Pennsylvanian backwoodsman on the excitement of ring hunts.

Tuan notes that, in the fourth century B.C., Mencius reported the burning of woodland to deprive dangerous animals of their lairs, and that burning continued through the millennia. Large-scale deforestation led the Ming to forbid over-cutting in 1580: forest grew back on the mountain sides but began to be cleared again in Qing times, despite another prohibition of logging in 1683. The ability to carve cropland out of forest did not succeed in permanently raising per capita incomes, but meant more and more people could be fed. It helped perpetuate the Imperial system.

Expansion was as much a leitmotif of history in China as in the West, though it is far less known. The Han Chinese spread into the warm, biologically rich, southern forests, a colonisation that Georg Borgstrom called one of humanity's greatest acts of ecological stupidity.[v] Timber was felled or burned on some 670 million acres, 28% of the land surface. 'Misfortune' might be better than stupidity, since it is not clear what alternative Chinese society had in the past. Men from the densely settled regions longed to subdue the forest as ardently as did any European peasant. They replicated the paddy-field countryside from which they came, knowing no other way to farm, just as settlers in the neo-Europes replicated as best they might the rain-fed farming systems of Old Europe.[vi]

During the Tang dynasty (A.D. 618–904) four river fish were added to carp as cultivated species. They helped to control malaria by eating mosquito larvae and thus, when the time came for the southward push, rendered large areas more habitable than they might have been. Fish not only supplied most of the non-vegetable protein in the diet but surplus fish were put on the fields as fertiliser — like the folk practice of East Anglia and New England. Dryland crops, fetched from the Americas as part of the Columbian Exchange, made it possible to colonise the once-forested southern uplands, eventually coming to supply perhaps 20% of total food production. Later, during the second half of the nineteenth century, settlement did recoil from the highest mountain sides. But little seems to be known of the ebb and flow of wildlife and vegetation associated with the entry and rebound of cultivation from marginal areas.

CHINA'S ENVIRONMENTAL PRESENT

Until the past few years, China treated self-sufficiency in food as sacrosanct. Crossing the bridge to dependence on imported food is significant in the history of any country, notwithstanding that some city states crossed over in the remote past. A suspicious polity like China has been reluctant to take the same step. The communist government fears that domestic instability might result from food shortages and is nervous about relying on shipping routes which might potentially be threatened by the American navy.

Nevertheless, rising urban incomes have meant rising demand for livestock feed and textile raw materials. Once China joined the WTO in 2001, imports soared. The United States became the world's largest exporter of soya beans and cotton and China became its top trading partner. These commodities are not as sensitive politically as cereals. China had kept self-sufficiency in grain production as a national goal but in 2010 did import maize from the United States. In 2012, there was finally an admission at the official level that the country needs to obtain maize, wheat and rice in the international market.

The previous insistence on self-sufficiency ignored the resource cost and the fact that the area of the average farm is only one acre, with productivity levels much lower than American benchmarks. Achieving a really big rise in farm productivity would require the emergence of a land market, so that holdings might be amalgamated and economies of scale reaped. Nevertheless the political implications of mass private land ownership are still shunned. In the circumstances it seems curious that the Chinese government has not found effective ways of slowing the conversion of farmland to urban and industrial uses, if only to assuage the discontent of dispossessed farmers.

Current Western commentary on the general Chinese environment is negative, for good reason. Of the 30 most polluted cities in the world, 20 are in China. Air and water pollution contribute to hundreds of thousands of deaths each year, racking up a huge medical bill and constituting another depressant on productivity. About 26% of the water in the 7 biggest rivers is unable to support animal life, and control measures are usually ignored. The pollution, erosion, deforestation and impoverished biota that can be observed are intensifications of millennia-old environmental degradation, hypertrophied by industrialisation and growth. Under Mao, talk of these ills was suppressed and the fellow-travellers who then visited the country were blind to them. They were first brought to Western notice in the 1980s by Vaclav Smil, who now has an army of imitators.[vii]

Ordinarily it might be predicted that public outrage at environmental degradation would work its way through political channels, but the state has managed to stifle the self-correction that the system needs, while for years citizens seemed willing to trade off pollution against continued economic growth. Highly skilled expatriates, important to China's economy, are less willing to put up with Beijing's air pollution (toxic smog nearly 40 times higher on some days than the levels considered healthy by the World Health Organisation), and there are fears of a sizeable exodus.[viii] Domestic Chinese incomes have risen far enough to embolden the expanding middle classes into pressing the authorities to remove noxious industrial works from their backyards. Tens of thousands of complaints and disturbances occur every year and the number is rising steeply. This may be a harbinger of wider activism leading to the political pluralism feared by Beijing, though the government has sometimes spiked its critics' guns by relocating offending plants more quickly than a democracy could manage. The government may be more willing to listen to expatriates' concerns, although so far it has continued to encourage car ownership and coal-fired power generation.

The Chinese leadership tries to save face by manipulating international opinion, succeeding in having one-third of the content cut out of a World Bank report on pollution on the grounds that the findings on premature deaths might cause 'social unrest'.[ix] Yet the deaths reported were fewer than appear on the website of China's own SEPA (State Environmental Protection Administration). SEPA calculates that 70% of the two million annual deaths from cancer are related to pollution.[x]

The negative externalities of China's growth spill over the country's borders.[xi] South Korea and Japan experience sandstorms because of desertification in China, acid rain hurts their forests and an explosion of the giant Nomura's jellyfish reduces Japanese catches of salmon and yellowtail. Nomura's jellyfish spawn along China's coast and the increase in their numbers is probably a result of nutrient run-off from farms and factories. These forms of damage fall within China's 'ecological shadow', where its activities cast a pall outside its own territory. To be fair, some of the following examples of harm-at-a-distance are stresses produced by the super-rapid growth of the whole of East Asia, blighting land and sea far around the globe. In principle, this is little different from the capacious maw of Victorian Britain; in practice, the excessive take of wildlife and plant products bears the signature of environmentally unconcerned Confucian culture. The East Asian scouring of distant sources has taken place faster than Western historical precedent: economic growth has been faster and data about resource availability, besides the technologies of acquisition, are more available. The change has come about recently and rapidly, Japan's growth having been marked only since about 1960, while China's opening to world trade in manufactured goods awaited the late 1970s.

Regional tastes are simply not the same as in the West. There is a home market to satisfy, with idiosyncratic uses for medicines and luxury foods. To satisfy these, obscure and endangered organisms are sought. East Asia's export industries also want raw materials of types that the West imports too, but the scale and rate of growth of Asian demand is unheralded. The most important are oil, natural gas, coal, iron ore and timber. Competition for these has become fierce, leading to the mineral exploration of remote parts of Africa and Latin America. Heavy investment occurs in weak states that find it hard to resist the sway of rich Asian countries. A new phase of geopolitics has opened. Even the developed economy of Australia has seen many of its eggs drop into the Chinese basket, as formerly happened when Japan began to pay attention to the southern hemisphere. The extent to which Australia has come to rely on continued growth in China is extreme and produces serious economic distortions; investment in the mining sector tends to crowd out investment elsewhere in the economy.

FOREST INDUSTRIES

Timber is one of the chief biological resources to enter world trade, which means unbridled deforestation in East Asia's southeast Asian hinterland.[xii] Or so it is usually said, together with statements to the effect that a total of one-half of Asia's forests has been cut and a series of countries have been intensively logged one after another. In several, only about one-fifth of old-growth forest remains standing. The Philippines had already turned from an exporter into a net timber importer by the 1990s. The most telling estimate may be that by the end of the twentieth-century Indonesia had only one-quarter of its old-growth forest remaining, yet this was half the total surviving in Asia and moreover held more species of plants and birds than the entire African continent. Japanese imports from some states have fallen but are still rising from others, the trade having recovered from a dip in prices during the Asian Crisis of 1997–1998.

Not all deforestation is to be attributed to logging. Much has been carried out by poor farmers who, when their societies are destabilised by incomers, cease their traditionally careful resource management and discount the future heavily. They clear-fell forests in a rush to obtain land for crops, including tree crops like rubber and palm oil. During the early 1950s, the government of China introduced rubber to Xishuangbanna in its tropical southwest. The intention was that rubber would be grown on large collective farms as a strategic resource, the need for which had been highlighted by the Korean War.[xiii] Xishuangbanna is a biodiversity hotspot, which, although it makes up only 0.2% of the country's area, holds as many as 36% of its bird species and large proportions of its mammals and plants.

In 1982, the Chinese government switched from collective farming to supporting individual entrepreneurs, encouraging them to plant rubber at low elevations and later on hill slopes. Farmers now had to pay cash for services formerly provided by the communes. They expanded the area under rubber plantations by 324% between 1988 and 2003. This was at the expense of forest and areas of former shifting cultivation, which had carried more varied wildlife than permanent crops. The standard of living has risen, which, coupled with a growing population, threatens further assaults on the environment. The government thinks of rubber, tea and sugar as 'green industries', yet this seems little more than a form of words: greener than factories, no doubt, but these crops make the land much poorer in species than previous uses. Similar changes are taking place across Southeast Asia.

We should pause at this point. Asia is designated the most damaged of continents, half of its forests and fish stocks having been lost and one-third of its land 'degraded' during the final 30 years of the twentieth century, but the calculation depends on how far back in time the baseline is set and even more whether it is credible.[xiv] Assertions about the felling of the world's tropical forests have been shown to be highly exaggerated.[xv] Estimates of the extent remaining, made within three or four years of one another during the 1970s, were most unsatisfactory; they varied by as much as 238%![xvi] A later study reports no net decline in tropical forest since 1970. This is the result of a winnowing of the United Nations data by Alan Grainger of the University of Leeds, who observes that it is unwise to aggregate often incommensurate national statistics.[xvii] Surveys by satellite ought to be more reliable. They show that not all is lost, although Brazil's own first satellite data exaggerated the loss of Amazonian rainforest. Asia, certainly, still contains some areas biologically so outstanding that 'extremely rich' would not suffice to describe them; they are amazing. Their survival is the more extraordinary in that some of them occur in the most populous urban areas, almost unbelievably including Hong Kong.[xviii]

The Special Administrative Region (Hong Kong is a region of the People's Republic of China) contains more species in most major plant and animal groups than the whole of the United Kingdom. There are 390 native species of tree and over 230 species of butterfly, while in all categories over 200 species new to science were discovered in the five years to 2002. For all the country's history of intensive land use, forests in the southwest of the People's Republic of China (PRC) are still able to supply the markets in Hong Kong with large numbers of wild birds, mammals, reptiles and amphibians.[xix] The natural productivity of the region's forests keeps Hong Kong's marts filled. China's remaining forests are very different from the bland uniformity of the rice paddies.

Despite their richness, habitats within the SAR itself remain threatened. Worse for long-term prospects may be the fact that biological hotspots no longer coincide; as a general rule the fragmenting of habitats is nearly as harmful as reducing their total area. Yet Hong Kong's inventory remains remarkable. No one, though, should permit this to disguise the losses inflicted by economic growth across East Asia as a whole. Moreover, The Conservancy Association of Hong Kong is less sanguine about the local situation than the *Biodiversity Survey* and more conscious of the struggle to retain and clean up the natural environment. A difference in emphasis is perhaps to be expected because specialists can find much more in the course of detailed research than amateur naturalists come across every day.

Scouring for resources in the forests of Guangdong and elsewhere in the PRC marks the externalising by Hong Kong of its search for wild foods and traditional medicines. This is part of the worldwide search for these things by all East Asian economies. To what extent is this sustainable? It is impossible to say for sure: the situation may be a little more sustainable than appears, thanks to the recuperative power of tropical environments. Some types of resources have withstood exploitation for many centuries, though populations rich enough to consume imported gourmet foods and quasi-medicines were smaller in the past. On the other hand, long-term consequences may be masked and the trade enabled to continue only by tapping ever-widening circles of supply, such as poaching bears for their paws and galls far away in Canada and the United States.

BIRDS' NEST SOUP

Consider the salival nests of two main species of cave swiftlets, which have been collected in Borneo for over 1,000 years, though not continuously, so that the Chinese (and Chinese-Americans too) may make birds' nest soup.[xx] This is a profitable business, run (it is alleged) by a tight nexus of brokers in Kowloon. Indonesia, which accounts for 70% of total production, was receiving US$200–250 million per annum for birds' nests in export earnings at the end of the twentieth century, and an estimated $226 million in 2009. About 100 tons were imported into Hong Kong in 2002. The trade has long persisted in what must be assumed to have been equilibrium, despite fragmented individual rights that might have been expected to frustrate conservation. The owners of the rock faces on which nests are found were Malays, whose rights tended to fragment because Islamic inheritance laws promote the sub-division of holdings. But small units are not very profitable individually, and this has assisted late-coming Chinese settlers, who possess some capital, in buying up most of the cave rights at Niah, Borneo.

Into this situation intruded the Suharto government of Indonesia.[xxi] The regime, with Suharto's own grandson conspicuously involved, seized the nest-harvesting rights and replaced them with an annual auction. It led rapidly to an excessive take. The story is not however the usual one told about land and resources in Southeast Asia, which stresses the poor definition of property rights and hence unbridled exploitation once prices or population start to grow: no 'tragedy of the commons' here, the rights were perfectly well understood and managed by custom. Suharto replaced the arrangements with an apparently logical commercialisation but which was really nothing to do with the more efficient allocation of resources, just a way for outside money to acquire rights behind a facade of legality.

Following the ousting of Suharto, there was a legal vacuum in which the looting of all non-timber forest resources began and actual fights took place over cave swiftlet nest sites. Outside companies that had bid successfully for harvesting rights became locked in conflict with reformers who wished to abandon the auction system and restore the rights to local families, who were supposed to be likely to return to safeguarding the goose that lays the golden egg. The matter could not, however, be resolved so simply. Despite the fact that every foot of rock containing nest sites had previously been owned, named and registered, the sheer growth of human population was already disturbing the equilibrium before Suharto's intrusion.

Birds' nest soup being one of the most lucrative animal products in the world, contention has not ceased. The nest sites attract poachers who lift the nests before the young have fledged, reportedly bringing about a steep decline in the number of birds.[xxii] This was countered on the market by fake substitutes, which forced the price down. At that stage, the Malaysian government had chips put in the nests to guarantee their authenticity.[xxiii] Furthermore, at various places along the coasts of Borneo and Northern Sumatra, the birds are being attracted to nest in custom-built concrete houses (rather as pigeons were farmed in dovecotes) by tape-recordings of their calls.

ENVIRONMENTAL PRESSURES

Nest houses for cave swiftlets may be an exception to prove the rule about the deleterious effects of East Asian tastes for wild foods; the growth of fish-farming is another. The pressures on wild resources are intense. Let's start with traditional diets. They feature rice, into which a few flakes and shreds of fish, chicken or egg are stirred. Animal protein was scarce in an agricultural system so devoted to maximising the output of rice for the human population that little land was allocated to pasture. The result was that East Asians have always been eager to consume animal protein in any form they can find. Very varied species of animals are sold in shops and restaurants in Hong Kong, Macao and Taipei, often killed in front of the customer in the traditional guarantee of freshness. The clientele is anything but wasteful: besides eating what to modern Western eyes is an exotic, even distasteful, range of species, it consumes almost every part of an animal. The blood of live snakes is drained and drunk in Taipei.

Couple this omnivorous proclivity with rising population, add one or two generations of rapid income growth, and the demand for protein swells like a bullfrog. Whereas the people of East Asia have acquired, embraced and improved manufactured goods once the preserve of Westerners, they have been slower to emulate Western diets. Or rather they have added items to their traditional cuisines — hence the maize imported to feed beef cattle in China, the import of Wagyu cattle bred in Australia to make Kobe beef and the elevation of Hong Kong, some decades since, to the middle rank of countries for heart attacks, thanks to adopting Western fast foods.

Japan gives indirect testimony to the adoption of Western lifestyles. Since the 1990s, the increase in garbage in Japanese cities has supported a big increase of crows. In Tokyo's parks, there was a total population of 7,000 crows in the late 1980s but 36,400 in 2001, a number supposedly cut to 18,200 by 2007, although the true figure was probably 150,000.[xxiv] Tokyo has been a fast food city ever since the Tokugawa shoguns, but modern waste has truly benefited the crows. At the same time, the demand for traditional forms of protein has not slowed. Instead sources have been sought further and further from East Asian shores.

Dismay at the poaching inspired by this search is thoroughly understandable. Nevertheless, to say, as Marc Reisner does in an otherwise gripping account of international poaching, that, 'in the midst of this rocketing economic growth, Asian society has remained a strange muddle of hypermodernism and atavism', lacks empathy.[xxv] He is right in principle to note the paradox that, 'multi-millionaire computer entrepreneurs in Korea still believe in the therapeutic properties of bear galls, a folk medicine reaching back four thousand years.' Yet the antiquity of the beliefs makes it unreasonable to expect them to be cast off overnight.

Change is slow and, in absolute terms, the demand for exotica is still increasing. Images of assaults on endangered species throughout Asia were on display in the BBC Wildlife Photography of the Year exhibition at Nature-in-Art, Twigworth, Gloucestershire, in March 2011.[xxvi] Particularly graphic were the giant snakes killed in Sumatra for the Chinese market. Black bears in Vietnam were shown being pumped for their bile, which is used in Chinese medicine. A Bear and Tiger Park in China was found to be involved in illegality, meat served in its restaurant having been shown by DNA analysis to be that of tiger! Every tank in Shanghai supermarkets contains a live turtle for the pot, fresh eels and frogs and even sharks on ice.

What is being observed might be called cultural lag, though this would imply that there is something especially reprehensible in East Asians clinging to traditional items of diet. This anthropological disdain is not helpful. Changed circumstances, i.e., economic growth, are precisely what make it possible for more and more traditional consumables to be afforded. Nor is this a uniquely Asian phenomenon. Even after the Second World War, English villagers ate every part of a pig except the squeak. In more prosperous times, redundant parts of the pig have returned as delicacies: tripe is offal yet it is served in expensive restaurants. Urban Westerners forget their own past and express horror at East Asian willingness to eat a comprehensive a range of cuts. They expect rising incomes to have brought about an instantaneous migration to Western dietary habits. After all, East Asians migrated to city living very fast, why could they not alter their dietary preferences as well?

We need to pause again. An equivalent of berating East Asians for retaining old dietary habits would be to criticise Westerners for eating the bubble-and-squeak or bread-and-butter pudding that have reappeared on fashionable menus. They are merely ways of using up leftovers. Western city dwellers forget how recent the shift to intensively produced food has been in their own societies. They forget the menagerie that once filled London markets like Leadenhall and how short of protein rural dwellers were. Cottagers ate what they could get, including 'vermin pie'. And all sorts of oddities are still considered treats by some Westerners, such as whelks, snails and frogs' legs.

Part of what is being observed is that income levels and underlying ecologies are particularly out of phase in the East, where economic growth is more recent than in the West. There are however signs of change. In Vietnam, the greediest consumer of endangered species, organisations such as the World Wildlife Fund have enlisted hundreds of restaurants in a campaign to reduce the consumption of wildlife products.[xxvii] And if that seems the result of outside intervention, educated young Asians are starting to spurn old cultural norms on their own account. The furore over dog meat in China suggests they may be on the brink of ceasing to eat it, as they are slowly ceasing to eat sharks' fins. In 2004, an animal rights group in Hong Kong shamed a supermarket chain owned by Hutchinson Whampoa in southern Guangdong into taking dog meat off its shelves.[xxviii] The chief campaigners against the cruel practice of slicing the fins off sharks and throwing the torsos back into the sea are themselves now Chinese.[xxix]

THE FISHING TRADES

Fish and other marine resources figure prominently among the items consumed. Take eels, which are catadromous, i.e., spend most of their time in fresh water but go out to sea to spawn. European eel populations might suffice indefinitely for the present level of home demand but the Chinese and Japanese market threatens to overwhelm them.[xxx] Glass eels (immature eels) are caught when they enter European rivers; most are sent to China, where they are raised to an edible size. Chinese buyers act as middlemen and processors. Since about 1986, Japanese demand has risen sharply, with about 60% of the supply arriving ready-grilled from China. The trade is partly surreptitious, with many sales unreported and reliable data hard to find: during the mid-1990s, Japanese buyers waited in the United States, cash in hand, by the river banks and it is therefore no surprise that market tallies and price data do not match. The share of eels sent to East Asia from Europe is equally uncertain but is considered large enough to have reduced stocks drastically.

Marine species have been fished intensively for years, although the fishing industry in the Pacific was somewhat eclipsed from about 1870 to 1920 because meat was then so cheap. Since the 1960s, Japan's fishing fleet, to say nothing of its whalers, has often been accused of over-fishing, though China's fleet has overtaken it as the largest in the world. The modern history of regional fisheries has been comprehensively analysed by John Butcher in *The Closing of the Frontier*, where he refers directly to the threat to less usual species by saying that, 'the emergence of a class of rich business people in the People's Republic of China renewed the centuries' old demand from China for the marine exotica of Southeast Asia'.[xxxi] Supply does not come only from Southeast Asia: Butcher includes a section called 'The Lure of Australian Waters', which have been a magnet for centuries, anciently attracting Chinese voyages to the north coast for trepang or *bêche de mer*. Nor is the demand for seafood purely Southeast Asian. The international web of traffic in the various products is complicated. The United States imports most of its own large and rising consumption of shrimp from Thailand, Indonesia and Vietnam, together with Ecuador.[xxxii]

Fishing industries grew fast from the late 1970s, responding to demand from Southeast Asia which was increasing at 2% per annum. What Butcher terms the marine frontier closed fast and biodiversity suffered, though biomass may have suffered more. Fortunately a ban on trawlers, for instance, in the Samar Sea (which is almost surrounded by the Philippines), enabled demersal, i.e., bottom-dwelling, species to recover quite rapidly. On the other hand, Butcher is unimpressed by what many authors regard as the most likely saviour of marine species, aquaculture, which has expanded enormously. Aquaculture may account for half the fish brought to market in the region but it destroys mangrove forests and hence the nurseries of fin fish.

A wide range of species has retreated in the face of the fishing onslaught. For illustrative purposes, they may be treated as more or less fungible: the toll taken of one may stand proxy for losses of another. About 83 species of whales, dolphins and porpoises lack any protection under international law. Dolphins and small whales are slaughtered according to an ancient (and now secretive) ritual in one Japanese village. Approximately 18,000 Dall's porpoises are also slaughtered in Japan every year. Some 70 million sharks are taken for their fins, which are sliced off to make sharks' fin soup. The National Parks Service of Galapagos confiscated 2,291 fins in January 2004, about 10 times the previous monthly average.[xxxiii] Feeding frenzy is a term applied to sharks but applies equally to the way in which humans abuse them. Vast numbers of sharks are said to die because they are accidentally caught by boats fishing for tuna or swordfish. Worldwide, several species of sharks are judged to be on the brink of extinction. The overall size of the catch is hard to assess, given that a total in the unhelpfully wide bracket of between 23 and 76 million sharks, or sharks' fins, is said to pass through the Hong Kong market every year, each fin fetching from $100 to $300.[xxxiv] The irony is that the fins are of low nutritive value.

Whale sharks, a gentle giant of a species, feed on plankton and krill and can grow up to 20 metres long, weighing 11 tonnes. In 1998, the world's largest known population was discovered by foreign fishermen off the Philippines, 400 kilometres south-east of Manila. Commercial fishermen descended on the area and Chinese-Filipino businessmen exported the carcasses to restaurants in Taiwan. Within a matter of *weeks* the whale sharks faced being wiped out.[xxxv] Commercial fishing was vastly more depleting of stocks than the take by local fishermen, who traditionally harpoon the migratory herds from small outrigger boats and for whom this is the only substantial source of income.

MALE ANXIETIES

Exotic foods constitute more than nutrition. They are often simultaneously elements in the region's ancient pharmacopeia, which is compounded of plant and animal materials (for status reasons, preferably the rarer ones) which happened originally to be at hand in village communities. They claim to rectify a great assortment of pathologies, with special attention to masculine anxieties about potency. Fear of impotence or male erectile dysfunction dominates the scene and means there has always been an expensive clutching at straws. Bezoar stones, which are hard, compacted masses of hair or vegetable fibre from the guts of monkeys, were anciently valued as aphrodisiacs and were traded from Borneo to China, while as recently as 2010 a speculative fever struck Kunming in south-west China over a caterpillar fungus thought to be an aphrodisiac. This raised its price close to that of gold.[xxxvi] Everywhere one looks, the demand for natural aphrodisiacs leads to ever more active searching and provokes conflicts among traditional hunters or fishermen, commercial enterprises and conservation bodies. The Galapagos provide an illustration, since the poor Ecuadorian fishermen insist on taking more sea cucumbers (*bêche de mer*) than should be harvested.

An appraisal by Professor Monique Simmonds of Kew Gardens is illuminating about the relative trustworthiness of plant materials employed in Chinese medicine (in the U.K., 500 species are traded in Chinese outlets).[xxxvii] Research so far carried out on the chemistry of half the species shows that little harm is done, which might perhaps be expected when the applications have been sieved empirically over the millennia. Cures relying on the body parts of animals are different: they depend for their presumed success on imputed behavioural attributes of the animals, like strength and prowess. Agreed, a placebo effect cannot be discounted, indeed is almost guaranteed to be the sole positive effect. In its customarily sardonic way, the *Economist* has dismissed the medical and aphrodisiacal effects as about as exciting as chewing your fingernail.

Traditional nostrums persist alongside high-tech medicine, even among those who can afford the latter. The advent of Viagra has not snuffed out the use of natural substances. Belt-and-braces appears to be the watchword and environmental costs are discounted. They are high. Rhinoceroses are especially threatened by demand from China and Vietnam. Powdered rhino horn is a favourite treatment for impotence, as are potions made from the remains of the masked palm-civet, an animal like a mongoose.[xxxviii] In late 2010, a time of record gold prices, rhino horn was selling for more than its weight in gold and over 260 had been killed so far that year.[xxxix]

Stag penises and three-penis wine (dog, stag and seal), as well as snake penis wine, are popular. In 1997, a Norwegian company sold its entire stock of 6,000 seal penises (an item few would expect to find in the international trade statistics) at $40 for the long ones (45 cm) and $15 to $20 for the short ones (20 cm) to a Singapore buyer, who took an option on all the seal penises the company could supply the following year.[xl] Seals, deer, dogs and many snakes are not under imminent threat, but rhinoceroses are seriously threatened by the quest for aphrodisiacs. About 90% of all rhinos are found in South Africa. The population of black rhinos was estimated at 100,000 in the 1960s but had fallen to 4,800 by 2011.[xli] White rhinos, which had been reduced to 50 a century ago, did recover hearteningly to 20,000 but they are threatened again by poaching and by the exploitation of loopholes in South Africa's hunting laws.

While one can discover dubious, often illegal, trade in an astonishing range of animals, the threat to the tiger is particularly worrying.[xlii] By the end of the twentieth century, only about 5,000 tigers were left in the world, mostly in India, compared with 100,000 a century earlier. The Bali, Caspian and Javanese sub-species were already extinct. Obtaining body parts for use in traditional remedies is the main aim of the kill. During the 1990s, staff of the London-based Environmental Investigation Agency visited pharmacies offering traditional Chinese medicines in Britain, Japan, the Netherlands and the United States. The 'products', so-called, were ground-up bones and organs used in tiger-penis wine and other purported remedies for various ills, notably impotence. Other parts pass as good luck charms, while tiger skins, teeth, claws and skulls are sold as ornaments.[xliii]

Trade in these items continues: 10 years into the twenty-first century, at most 3,000 tigers remained at large. It is one thing to accept that the actions of individual consumers are culturally understandable but the Chinese government has compromised itself by advocating lifting the ban on the domestic sale of tiger parts.[xliv] In 2007, there were tiger farms in China containing a total of about 4,000 animals, supposedly bred for release in the wild. The problem is that tigers are far cheaper to poach than to rear and the sale of parts from farmed tigers may camouflage illegal transactions.[xlv] Although Premier Wen Jiabao promised in November 2010 that China would combat poaching and the trade and smuggling of tiger products, the country has stealthily resumed trading in tiger and leopard skins and declines to state how many have been sold. The Chinese government does not even monitor the trade for itself but relies on the findings of NGOs.

ORNAMENTS

Ornaments are yet another traditional category of demand and when made of ivory have a major impact. Ivory carving has a long history in the region. From prehistoric times until the eighteenth century, the solid casque of the endangered Helmeted Hornbill was a recognised tribute piece, forming a valuable export from Borneo to China, where it was carved into ornaments more costly than jade.[xlvi] Walrus ivory is another prized item. But chief among species harmed is the African elephant, victim of the swelling trade in carved ivory in the countries of East and Southeast Asia, and India, all of which were historically exporters but with the advent of rapid economic growth have become end consumers.

Following the international banning of sales of ivory from 1989, elephant populations did gradually recover, but CITES (the Convention on Trade in Endangered Species) began to make exceptions. One-off sales were permitted in 1997, 2002 and 2008. China and Japan were able in 2008 to bid collusively for ivory from the official stockpiles of four sub-Saharan countries, containing tusks that had been confiscated from poachers or were from animals that had died of natural causes. By August of the following year, 11 tonnes of illegal African ivory had been impounded en route to China, giving credence to the suspicion that lawful sales camouflage unlawful ones. Chinese buyers were active at illegal markets in Addis Ababa and apparently found it easy to smuggle items like bundles of ivory chopsticks through customs. Many Chinese smugglers were arrested at African airports early in 2011, but two had to be released when they flashed diplomatic passports.[xlvii] Another market (ignored by CITES) is the Philippines.[xlviii] It may be significant that the Vatican has not signed the CITES treaty and that Manila has numerous ivory shops and carvers supplying icons to the Roman Catholic Church. In Thailand, Buddhism provides a comparable market.

An estimate, too broad to be very useful, is that between 4,900 and 12,000 elephants were being killed every year to feed African, Dubai, Yemeni, Thai, Vietnamese and Chinese carvers with ivory priced in May 2011 at US$1,200 per kilogram.[xlix] All seemed not quite lost, however, in that from a low of 500,000 in the 1980s the total African elephant population climbed to over 600,000, since its expansion in southern Africa (with a growth rate of 4% per annum) more than offset a decrease in Central and West Africa.

Yet a Damocles Sword hangs over the elephant, given that political campaigns in favour of ivory sales are as unscrupulous as those to promote commercial whaling. In 2011, CITES brusquely ousted all NGOs from its meeting in Geneva. NGOs are unaccountable and may be irresponsible, for instance, promoting scares about genetically modified food, but despite these defects, CITES' lack of transparency bodes ill. A little more promising are advertising campaigns to persuade rich Chinese not to buy ivory, which they often do not understand comes from slaughtered elephants.[l] But not much more promising. The recovery of elephant numbers after 1990 has stalled, with an estimated 25,000 killed in 2011 and almost certainly more in 2012. The viability of Africa's elephant population is threatened. Demand is 'traditional', like India's demand for gold (also at a peak in 2012), and therefore may take a long time to weaken.

The litany of commodities entering trade between the outer world and East Asia runs on and on. Some trades are old, if not truly ancient. Trafficking from French Canada to China in ginseng, a plant root with the usual supposedly aphrodisiacal properties, began in the early eighteenth century. The Qing Emperors built the 'Willow Palisade' to protect their Manchurian homeland, thus keeping out the ginseng hunters among the Han Chinese they ruled. Late Colonial America and the early United States found the market in China wide open. Likewise, in the early nineteenth century, American sealers were soon shipping to China the pelts of fur seals taken in the southern hemisphere.

Still other commodities testify to modern communications and the drawing power of the region's recent wealth. Japanese demand revived the nineteenth-century gathering of fresh-water mussels from the Mississippi, so that, as early as 1966, the export value was worth nine million dollars (in current dollars): 'hoovering up' and 'dramatic decline' are expressions that come to mind.[li] Trade worth perhaps US$500 million per annum features 50 million product items manufactured from reptiles, chiefly leather from the hides of American alligators and Nile crocodiles. Some of these are destined for French and Italian handbag and shoe factories, which have outlets in China. The galls and paws of black bears illegally slaughtered in British Columbia are laundered through Ontario (where protection extends only to items from the province itself) and Quebec (where protection is absent); they are consumed in China and Vietnam, as well as by people of Asian descent now residing in North America.[lii]

PROSPECTS

The East Asian market is a whirlpool sucking down the natural world. Much of the trade has no scientific basis, the medical value of the products being at best unproven and items like ivory being used for needless, frivolous purposes. Despite this it is carried on with the connivance of governments. On the other hand, it is possible, not so much to cry wolf, as to lament too loudly. Habitat change is still the more potent force. The poaching of endangered species is unlovely but forest clearance has more impact.

Yet something contradictory is occurring: while poor populations and logging firms carve their way into forest land, a large outflow of population from rural areas is taking place. Agricultural intensification means that a sizeable fraction of population growth is being fed from a smaller area than would have been needed had farming technology remained at the 1961 level. Not all is gloom. Two billion acres of secondary forest have grown up where farmers have left for the city and are almost as rich in biodiversity as primary tropical forest.[liii] An estimate referring to Panama and other tropical countries is that for every hectare of rainforest cleared, 50 hectares of new forest are growing on land formerly farmed or logged. Agricultural intensification has saved 44% of the planet for wilderness; from the environment's point of view, this is the best thing that could have happened, surpassing anything the conservation movement could achieve. The largest case study, of Puerto Rico, shows that despite losing 99% of the original forest, there are now *more* species of birds than before.[liv] Dudgeon and Corlett point out that forest in the Hong Kong SAR is expanding and maturing, so that it may be possible to reintroduce some of the species previously driven out, the forest pheasant for instance.[lv]

Environmental problems within the PRC are nowadays widely reported.[lvi] That China is partly externalising its problems is clear: an example is the fact that its forests are being conserved by virtue of exporting disafforestation. Obtaining accurate measures of China's domestic forestry industries is difficult and probably impossible. In 2011, the two biggest groups were accused of fraud and in June the biggest of all, Toronto-listed Sino Forest, lost over 80% of its value.[lvii] Trading in the company has been suspended. Industry data on tree growth are lacking, making it difficult to forecast output. There is no central land registry and forests in China are owned by collectives run by local government officials who are not fully under Beijing's control.

Two main forces keep up Asia's pressure on other parts of the world: firstly, the expanding economy's demand for raw materials and luxury items, which is unlikely to ease much while rapid economic growth continues. Damaging as this voracious consumption is in so many respects, there is no fundamental case against it: the Chinese people have every right to try to raise their standard of living. The second force (common to the East Asian countries) is more problematic, the persistent consumption of biological resources that either cannot be shown to fulfil their promise — aphrodisiacs again — or are pointless luxuries. The historical origins are plain enough but their ghostly survival into the modern world is too likely to extinguish entire animal species, tigers, elephants and rhinoceroses, to name only the most conspicuous. Whether rising prices, better education and the adoption of substitute commodities can curb this before it is too late is possible but not very likely. Modernity has overwhelmed East Asia without fully replacing medievalism.

ENDNOTES

[i] Yet it was in Europe that 300 conventions for the humane treatment of the sick and wounded were signed in the centuries after the first such at Tournai in 1581; J. Magill, *The Red Cross: The Idea and its Development* (London: Cassell, 1926), p. 10.

[ii] E. Jones, *Cultures Merging: A Historical and Economic Critique of Culture* (Princeton: Princeton University Press, 2006).

[iii] J. Stewart-Smith, *In the Shadow of Fujisan: Japan and its Wildlife* (Harmondsworth, Middlesex: Viking, 1987), p. 44.

[iv] Y.-f. Tuan, 'Discrepancies between environmental attitude and behaviour: Examples from Europe and China,' *Canadian Geographer*, **12** (1968), pp. 176–191; Y.-f. Tuan, 'Our treatment of the environment in ideal and actuality,' *American Scientist*, **58** (1970), pp. 244–249.

[v] G. Borgstrom, *The Hungry Planet* (New York: Collier Books, 2nd revised edn., 1972), p. 106.

[vi] E. L. Jones, *The European Miracle* (Cambridge: Cambridge University Press, 3rd edn., 2003), pp. 213–219.

[vii] V. Smil, *The Bad Earth: Environmental Degradation in China* (Armonk, New York: M. E. Sharpe, 1984).

[viii] *Financial Times*, 2 April 2013.

[ix] *The Economist*, 3 July 2007.

[x] *The Economist*, 26 January 2008.

[xi] *The Economist*, 31 March 2007.

[xii] E. Jones, 'The environmental impact of growth and poverty in East Asia,' *The Melbourne Review*, **3** (2) (2007), pp. 80–85; E. Jones, 'The history of biological exploitation on the Pacific Rim,' in I. Kaur and N. Singh (eds.), *The Oxford Handbook of the Economies of the Pacific Rim* (forthcoming November 2013).

[xiii] W. Li *et al.*, 'Environmental and socioeconomic impacts of increasing rubber plantations in Menglun Township, Southwest China,' *Mountain Research and Development*, **26** (3) (2006), pp. 245–253; J. Fox, 'Crossing borders, Changing landscapes: land-use dynamics in the Golden Triangle,' *Asia Pacific Issues*, No. 92 (December) (2009).

[xiv] C. Foy *et al.*, *The Future of Asia in the World Economy* (Paris: OECD, 1998), p. 197.

[xv] B. Lomborg, *The Skeptical Environmentalist* (Cambridge: Cambridge University Press, 2001), pp. 113, 117.

[xvi] In J. L. Simon (ed.), *The State of Humanity* (Oxford: Blackwell, 1995), p. 337.

[xvii] University of Leeds report by A. Grainger, 'Difficulties in tracking long-term global trends in tropical forest area,' *Proceedings of the U.S. National Academy of Sciences*, 8 January 2008.

[xviii] D. Dudgeon and R. Corlett (eds.), *Biodiversity Survey of Hong Kong* (University of Hong Kong, 2002), web version.

[xix] M. A. Webster, 'Hong Kong's trade in wildlife,' *Biological Conservation*, **8** (1975), pp. 203–211.

[xx] B. E. Smythies, *The Birds of Borneo* (Edinburgh: Oliver & Boyd, 1968), pp. 29–31; *BBC News Asia-Pacific*, 21 January 2011.

[xxi] *Far Eastern Economic Review*, 21 January 1999.
[xxii] *National Geographic News*, 28 October 2010.
[xxiii] *BBC World Service*, 17 July 2012.
[xxiv] *New York Times/Observer*, 18 May 2008.
[xxv] M. Reisner, *Game Wars* (New York: Penguin Books, 1992), p. 82.
[xxvi] Noteworthy images by Mark Leong are posted on Google Images and his own website. See also the illustrated interview by Alison Zavos, 'Mark Leong, Beijing,' *Feature Shoot*, 22 February 2010.
[xxvii] *Vietnam News*, 10 February 2010; *New Scientist*, 28 July 2012.
[xxviii] *The Economist*, 14 January 2006.
[xxix] *Financial Times*, 25 August 2011.
[xxx] R. Schweid, *Eel* (London: Reaktion Books, 2009), pp. 129–142.
[xxxi] J. G. Butcher, *The Closing of the Frontier: A History of the Marine Fisheries of Southeast Asia c.1850–2000* (Singapore: Institute of Southeast Asian Studies, 2004), p. 234.
[xxxii] *Financial Times*, 9 June 2010.
[xxxiii] *Financial Times*, 4 May 2004.
[xxxiv] *Financial Times*, 18 February 2008.
[xxxv] *The Australian*, 14 May 1998.
[xxxvi] T. Harrisson, *World Within: A Borneo Story* (London: The Cresset Press, 1959), p. 31. *The Economist*, 20 November 2010.
[xxxvii] M. Simmonds, Lecture on 'Plants for Life', Cirencester Science and Technology Society, 11 June 2008.
[xxxviii] *The Economist*, 30 May 1998.
[xxxix] *The Economist*, 17 August 2011.
[xl] *The Independent*, 11 November 1997.
[xli] *Time*, 13 June 2011.
[xlii] Yet no U.K. representative attended the International Tiger Conservation Forum in St Petersburg in 2010 because of an 'austerity drive' — this despite governmental waste in so many other directions.
[xliii] *The Independent*, 11 November 1997.
[xliv] *Financial Times*, 21 May 2007.
[xlv] Environmental Investigation Agency website.
[xlvi] Smythies, *Birds of Borneo*, p. 29.
[xlvii] *The Economist*, 21 May 2011.
[xlviii] B. Christy, 'Ivory worship,' *National Geographic*, **222** (4) (2012), pp. 28–61.
[xlix] *Financial Times Magazine*, 22 August 2009.
[l] *The Economist*, 3 November 2012.
[li] J. Madson, *Up On the River* (New York: Penguin, 1985), p. 90.
[lii] *The Economist*, 27 April 1996.

[liii] M. Ridley, *The Rational Optimist* (London: Fourth Estate, 2010), p. 144; *New York Times/Observer*, 15 February 2009. Note, too, that tropical deforestation between 1985 and 1995 was offset to a degree by an increase in the volume and area of timber grown in temperate realms. Simon, *State of Humanity,* p. 337.

[liv] Cited by B. Lomborg, *Skeptical Environmentalist*, p. 254.

[lv] D. Dudgeon and R. Corlett (eds.), *Biodiversity Survey.*

[lvi] A concise summary is in *Nature*, **435** (2005).

[lvii] *Financial Times*, 22 June 2011.

14. THE MODERN EXPANSION OF AGRICULTURE

> The impending quantum change in commodity prices is perhaps the most important economic event since the Industrial Revolution.
>
> *Jeremy Grantham*, Global Asset Manager (GMO), May 2011

In the language of public affairs, the most chilling phrase is said to be, 'it's different this time'. At every steep slide in the stock market, out come the Marxists to celebrate that this time capitalism really has reached the end of the road. And at every major spike in food prices, the neo-Malthusians emerge to proclaim that population growth has finally outstripped food supplies, this time for good. The food price spike of 2008–2009 produced this response. Influential opinion is convinced that price spikes are caused by speculators buying stocks of food and holding them against further rises. Blaming speculators predominates among conspiracy theorists in Europe, especially in France. They discount drought, rising demand or rising costs of production. In truth, despite being painful and even fatal for some of the world's poor, market prices do not reflect the supply situation fully. Most crops in less-developed countries are grown and eaten locally, and too small a fraction is internationally traded for market prices to represent reality.

SUPPLIES

Between 1862 and 2010, through decades of industrialisation and prodigiously growing population, *The Economist's* world commodity price index halved. This was a measure of advancing methods of production and distribution and, in the case of minerals, of fresh geological discoveries. Even for 'soft' commodities like foodstuffs, production has often relocated or become concentrated, presenting economies of scale and scope. Excluding the two World Wars, the world has become accustomed to slowly falling commodity prices, albeit with a saw-tooth edge reflecting year-to-year fluctuations in harvest weather.

The rupturing of this comfortable trend in 2007 to 2009 was more serious. Wars could not be blamed; they were vicious but not big enough to disrupt supply chains. The Malthusians reappeared from the shadows, borrowing Death's scythe and announcing a 'super-cycle' in which demand for foodstuffs would outstrip supply for decades, if not for ever. They said that the usual response of farmers curbing prices within a couple of years by planting a greater acreage, as happened in the scare of the mid-1990s, would be overwhelmed for one or both of two reasons: first, the technical means of raising output on existing land or finding enough extra arable was terminally diminishing. This was a sort of End of (Agricultural) History thesis. Secondly, demand would prove insatiable, because the inhabitants of China, India and other emerging markets were prospering and eating more meat, requiring ever more land to be switched to raising feed for livestock.

The super-cycle thesis envisaged unstoppable growth pressing against inelastic barriers. Economic growth was projected into the far future, without recognising the anomaly that this in itself presupposes sufficient food. The price spikes of 2007 to 2009 gave credence to a vision of tight supplies and famines. Newspapers that had previously ignored agriculture as too dull for words quickly filled with crop reports. In emerging markets, there was evident cause for fears about prices, the first signs being the 'tortilla riots' in Mexico in 2006. In 2009 and 2010, food riots were reported from over 60 countries. The timing of the Arab Spring in 2011 was attributed to rising cereal prices, especially in Egypt, where domestic food production was falling.[i] Egypt has to import 40% of its foodstuffs, 60% of the staple, wheat. Ration cards were extended to include rice and sugar.

Fear of popular unrest led several countries to impose bans on the export of grain, sugar and cotton. This is permissible under international law whenever supplies are deemed crucial to an exporting country's domestic market — essentially a loophole because national governments can decide for themselves when it applies. Bans exacerbate the difficulty because they reduce the incentive to extend plantings, for instance Russian farmers switched to oilseed because it was not included in the state's ban on exporting cereals.

Feverish excitement surrounded each rise in prices, though the benchmarks were scarcely consistent — 'higher than for thirty years' or whatever comparison came to mind. Investment advisers began to recommend that portfolios emphasise commodities and turned their clients' attention to companies making farm machinery or fertilisers. Mergers and acquisitions increased among fertiliser firms. The fever reached its height, or rather depth, in January, 2011, when an advertisement for spread betting on food prices appeared on Liverpool Street station in London. Alternatively, the height of hyperbole came in May 2011, in the form of Jeremy Grantham's incautious comparison with the Industrial Revolution, given as this chapter's heading quote.[ii]

Recession in the developed world then led to collapse of the commodity boom, and the daily papers sank back to their ordinary insalubrious fare. The question was left hanging as to whether the spikes had been early warnings of a break of trend that would soon resume. To be fair to Jeremy Grantham, he had hedged his bets by forecasting that, as the Chinese economy slowed, the wave of interest in commodities would come to a halt, though opining that the pause would be temporary.[iii] Economic growth continued to make the poor populations of the emerging markets richer and keener to eat more protein. Unfortunately, agricultural supply is relatively inelastic.

Running through reports on the major farm commodities in world trade, it is possible to discern that the price of rice had actually stayed more or less stable. This helped to insulate many of the poor in Asia from the worst consequences, although only 5% to 7% of the output of rice enters world trade. Other grains were not so stable. Corn (maize) output was high but not enough to satisfy the new ethanol market and human consumption simultaneously. The wheat situation was insecure: Argentinian controls led growers there to respond by reducing the acreage sown; India's storage facilities, often just tarpaulin sheets, were inadequate and stocks rotted; 30% of Brazil's grain is said not to reach the market because of poor transport; while in East Africa and the Middle East, wheat rust had been spreading since 1999. And the price of oats rose because rain in Canada (the largest exporter, with 80% of the world market) meant fields went unsown.

Because of the high cost of feed grains, internationally traded meat soared in price. The U.S. Beef Association admitted it had been caught flat-footed. A former source of meat, the backyard farming of cows, sheep, pigs and poultry in Asia was contracting as families began the more flexible practice of buying their meat in the markets. In Australia, the largest sheep-producing country, the national flock fell by 56% between 1990 and 2008, so that despite a hasty switch in emphasis from wool to meat, the output of mutton and lamb was continuing to shrink. In the other main producers, sheep numbers were down by 40% to 50% (80% in South Africa). In 2011, global wool production was at an 85-year low. Textile fibres as a whole were in short supply: the price of silk, of which China is the prime producer, rose because of urbanisation in the zone around Shanghai previously devoted to growing mulberries. As for cotton, its acreage was at a low because the increased use of artificial fibres had led growers to switch to crops like soya beans. By November 2010, cotton was trading at its highest price for 140 years.

Tropical crops also fared badly — but then displayed a sudden and unexpected recovery. Perhaps there had been no permanent reversal of trend in soft commodity prices after all. Perhaps things would not be different this time. Consider cocoa: prices had risen alarmingly because of civil war in Ivory Coast, which normally produces 40% of the world crop. The trees were aging and were owned by smallholders without the resources for replanting. Prices reached a 33-year high in 2010, but the rains came, supply soared, and in two weeks they dropped by one third. The correction was supported when new countries were drawn into the world cocoa trade by the anticipation of high returns. Sugar behaved not dissimilarly, having reached twice its 1974 price by 2010, but falling by 23% in just two days that November!

Food and fibres share a vulnerability to inclement weather but the shocks are not correlated. Detecting the overall supply trend is difficult, it being too easy to string together reports of steeply rising prices for one commodity or another, in one region or another, and construct out of the whole what seems to be a universal trend. Production has increasingly concentrated geographically, but not so much as to abolish the portfolio insurance of dispersed locations. Too much liquidity in the system seems more likely than the weather to have pushed prices upwards. The short-run reporting of crop failures, which are seldom total, encourages confirmation bias among the doomsters, but trade tends to offset a deficit in any one area. Globalisation clearly has advantages.

Similarly, apprehensions about a 'global pollination crisis' inspired a survey which found that the increase in the global managed population of honey bees was not the result of biological forces but depended on growing demand for, and trade in, high-value foods and hence for pollination services for the relevant crops.[iv] Nevertheless, the authors of the survey could not refrain from postulating some future stress on global pollination capacity; bows to negativism of that type are endemic in the literature. From the point of view of the natural environment, their more pertinent observation was that honey-bees are vigorous and 'steal' pollen from native plants, reducing their seed production. As for journalists, their patron saint is Chicken Little. Once they sense a scare story, they are disinclined to make space for good news. Modern communications make it easier than ever to join up the dots, as it were, and draw the looming ogre of global famine. This tendency dominated the 2008 to 2010 period, until the recovery of the cocoa and sugar crops at the end of 2010.

International commodity trade is largely in the hands of a small group of companies, not that this is new: it was said in 1898 that Central American countries were each 'ruled by a firm of coffee merchants in New York city, or by a German railroad company, or by a line of coasting steamers, or by a great trading house, with headquarters in Berlin, or London, or Bordeaux...'[v] At any period, trading companies need to have as much of a quantitative grasp of the variables as possible, though this is troublesome since the data are poor. Only in the United States, where the USDA pays annual visits to 11,000 plots and interviews 75,000 farmers, is much known about supply, demand and crop inventories.[vi] Better information from elsewhere might be expected to dampen price fluctuations, but many countries lack the political will or technical means to collect agricultural data, let alone disseminate it; China considers the size of inventories to be a state secret. When rising prices are intensified by a lack of storage or transport there is sure to be delay before the situation can be remedied. Substitutions among foodstuffs are therefore made, although many peoples' tastes are rather invariant to price, surprisingly given the fashion among the middle classes for eating foreign cuisines in restaurants.

The narrative about a super-cycle was unravelled by the supply response. Admittedly, the underlying increases of population and demand in emerging markets give no grounds for complacency. Regardless of any supposedly approaching biological limits, mundane considerations of infrastructural constraints and the protectionism stemming from 'resource nationalism' hamper productivity, as do populist policies like the European Union's resistance to GM crops. The resistance is scientifically unfounded in view of the 'trillion meal experiment' with GM elsewhere in the world. The chairman of Nestle, the world's biggest food company, hit out at 'well-fed activists' whose hostility to new food technologies is holding back agricultural productivity but opinions like his are seldom much publicised.[vii] Nevertheless, despite the depressants, coincident trends work in good as well as bad directions and there was at least a temporary reprieve for consumers, thanks to big harvests over much of the world in 2011. By October, India lifted its bans on wheat and rice exports.

Yet at the end of that year, Russia was mooting a re-imposition of its ban on exporting wheat, sparking another upturn in prices. Among governments concerned with national interests, and among the companies they favoured, a 'battle for acreage' continued. It underwrote the farmland grab in Africa, Latin America and parts of the Former Soviet Union by East Asian countries and the Gulf States. Western investors did not stand entirely aloof; the supremely cautious American pension fund, TIAA/CREF, invested US$2 billion in farmland in the United States, Australia and Brazil.[viii] About 54 other investment funds have a total of $7.44 billion agricultural investments around the world. Less run-of-the-mill expedients are being canvassed, with talk of vertical farms many storeys high, which would have advantages if the cost of artificial lighting did not prove prohibitive. Energy is a problem: Thanet Earth, which grows 15% of the British salad crop in 90 hectares of single-storey greenhouses, is obliged to run its own power station.[ix]

Panic about the possibility — some say the certainty — of a Malthusian future has inspired a literature and politics of food security. Along the perennial lines whereby anti-free trade arguments continually re-surface in new dress, much of the campaign is scarcely disguised Protectionism. About the Millennium, 'Multifunctionality' was popular as a term and policy, but food security is the newer label.[x] The implication is that each nation state ought to produce everything, as if everywhere possesses comparative advantage in all products. The thesis has a sinister history. A Nazi agronomist, Herbert Backe, persuaded Hitler that Germany must be self-sufficient in order to win the war and (besides deliberately starving unimaginable numbers of eastern Europeans) needed to seize another 7 or 8 million hectares.[xvii] This was part of the motive for invading Russia.

Wherever one looks, there is an anti-trade element in the campaign for food security and an absence of faith in market mechanisms. Yaneer Bar-Yam of the New England Complex Systems Institute, interviewed on how to moderate food prices, advanced the nonsensically self-defeating claim that, 'investors can help by avoiding investing in commodity markets for profit'.[xii] National governments incline to behave opportunistically in hopes of political support: in 2011 Thailand bought rice from its farmers at above market rates but in 2012 was unable to sell the stockpile without losing the equivalent of 5% of GDP. It had been undercut by record crops in India, the Philippines and China.

Some proposals sparked by concerns about food security recommend research aimed at raising output in general and some extol policies to ensure that, if others starve, at least their own nation will not be among them. Against this, the collective of rich nations, the G20, proclaims that 'strong global governance' is necessary, notwithstanding the world's feeble record in this respect — for instance FEWS NET, the Famine Early Warning Systems Network set up after the Ethiopian famine of 1984–1985, failed to mobilise a significant international response to the Horn of Africa famine of 2011.[xiii] International co-operation may sometimes succeed, but one can have little faith that stocks will be freely redistributed during severe shortages. Nor can one have much faith in the willingness of developed countries to allocate food sensibly or humanely, given that even in the United States 48.8 million people experienced or came close to hunger in 2010, an increase of 12 million since 2007. The current annual cost of this to the American economy is estimated at $6.4 billion.[xiv] This artificial scarcity is the other side of the obesity epidemic. Both problems are crying out for political solutions and have little to do with agriculture's physical capacity.

LAND ACQUISITION

The years when food prices spiked have seen energetic efforts by some countries to acquire overseas farmland, the 'farmland grab'. It is the equivalent of early modern Europe's acquisition of 'ghost acreage' overseas. What this does, so to speak, is to expand offshore the effective area of the home countries.[xv] In the modern world, the active countries are East Asian or Middle Eastern, and the target areas are chiefly in Africa, Latin America, parts of the Former Soviet Union and even Serbia. Most schemes aim at growing cereals but the role of palm oil should also be noted since the increased consumption of processed foods in India and China is expanding demand for edible oils. Asian producers who are running out of suitable land are starting to compete for acreage in Equatorial Africa; they are acquiring land or taking leases on huge areas in at least seven African countries. The costs of developing plantations and installing mills to crush the fruit for palm oil are formidable but the profits to date have been so large they are affordable.

According to the World Bank, land deals for all purposes involved fewer than 4 million hectares in 2008 but 45 million hectares in 2010. The Third Future Directions International Food and Water Crisis Workshop in March 2011 estimated that the world's cultivated area (then two billion hectares) would need to increase by 0.5 to 1 billion hectares to provide sufficient food for the global population anticipated by 2050.[xvi] That would be a most formidable increase but the workshop thought it could be achieved, there still being abundant unused or underused land in Africa, Brazil, Argentina, Russia, the Ukraine and Romania. These are the parts of the world to watch in terms of environmental impacts from putting fresh land under the plough. Presumably land left unused as people moved to the city will have to be re-used, negating a change that has been quietly offsetting deforestation.

The Institute of Development Studies at Sussex University examined over 100 deals and was mostly damning about the terms received by the countries ceding their land, though the details of most deals remain secret.[xvii] Almost 80 million hectares were subject to negotiations with foreign investors — more than the combined area of Britain, France, Germany and Italy. This great enclosure movement is continuing. Contracts were usually vague, local elites were being suborned and foreigners were carving out enclaves for their own farmers. In short, positive spill-overs to the host countries seldom eventuated and customary law was hardly safeguarding local inhabitants.

East Asian countries are prominent movers. The South Korean *chaebol*, Daewoo, expects to have to pay nothing to produce maize and palm oil on an area of Madagascar half the size of Belgium, taking a 99-year lease on 1.3 million hectares. This is intended to provide half of South Korea's maize imports (the country is the fourth largest importer). The sole *quid pro quo* will be work for the locals. Environmentally, the proposition is threatening, since Madagascar is one of the two leading countries in the world for endemic families of birds, New Zealand being the other.

Densely populated countries are targeting less dense ones but this is not the entire sub-text. There are examples where the process involves trying to raise productivity on the existing cultivated area. For instance, Singapore aims to establish a 'Super Farm' in northeast China, the 'China Jilin Modern Agricultural Food Zone' of 1,450 square kilometres, 200 times the extent of Singapore's own farm land. The project is to run for 15 years and involve an investment of $18.4 billion. At the same time, Chinese farmers are making sparsely populated regions of central Russia bloom.

Japan, on the other hand, eschewed land acquisition, not wishing to be criticised for investing in parts of Africa where the local inhabitants are short of food.[xviii] Instead, Japan proposes to invest in upgrading infrastructure, technology and the means of distribution. The aim is to support Japanese companies wishing to engage in international trade in foodstuffs, something that was already happening by 2009. With the exceptions of discussions with West Papua, and one Japanese firm that has leased 600,000 hectares to grow biofuel in the Philippines, there is to be no purchasing or leasing of overseas land.

For densely populated East Asian countries, the temptation to establish 'food security' via foreign agricultural land is obviously considerable. This makes Japan's self-denying ordinance and reliance on the market the more remarkable: it has fewer than 5 million hectares of arable land, would need 17 million hectares to feed itself at current nutritional levels and is the world's largest grain importer. China is similarly short of cultivable land relative to its population and, although remote sensing in 1998 showed previous official statistics had been too pessimistic, it continues to lose land to urbanisation. Farm acreages are tiny and potential economies of scale are not reaped. Whereas in the United States, the top three producers of pork supply 80% of the market, the top three in China supply fewer than 10%. Additionally, because China is at a later stage of economic development than Japan, it is experiencing the full force of Engel's Law, whereby the rising incomes of its people mean that a smaller fraction is spent on food altogether and of that fraction a higher proportion is spent on protein. As we have noted, meat, which (besides fish) is what protein mostly means in this connection, is very demanding of cereals for animal feed. Fish farming too is a heavy user of grain.

China entered world trade in foodstuffs only after joining the WTO, and then tentatively. As far as the PRC is concerned, food security did not suggest trade until very recently but rather trying to command supplies from what to all intents and purposes are colonial enclaves. Settlers from China had been trickling into Africa for some time but the process was made a priority in 2006 and an investment of $5 billion was committed in 2008. Ten agricultural demonstration centres were promised, though in practice it is the Chinese incomers who do the most vigorous farming. This is plain across the face of the land and tends to cause friction with local inhabitants. Agricultural projects remain 'shrouded in mystery'.[xix] Circumstances alter all the time, however; in 2012, China consciously moved away from its mantra of self-sufficiency and declared that in future it would rely on world markets for staple crops.

Arab countries — Saudi Arabia, United Arab Emirates, Kuwait and Qatar — also planned to buy or lease land to grow cereals for their own use.[xx] Saudi Arabia was already intending to phase out domestic wheat production because water is so scarce. But as prices began to subside the urgency faded and few projects really got started. Nevertheless,

Saudi Arabia was still considering investments in Ethiopia, Sudan, Ukraine, Cambodia, Vietnam, the Philippines, Turkey and Egypt. This implies a complicated overlap around the world among investing and host countries.

In 2008, Kuwait, Qatar, South Korea and the Philippines were all interested in schemes by which Cambodia, with an uncultivated area of 3.5 million hectares, hoped to attract $3 billion of investment in return for land concessions. Cambodia was also talking to Indonesia about ceding land for biofuel crops. Indian investors toured Ethiopia, Tanzania and Uganda seeking land to grow a variety of crops for their home market. They planned to spend $2.5 billion on buying or renting land, a sizeable investment in areas of significant political risk. Meanwhile, other Indian investors were trying to enlist their government's support for projects in Latin America.

Everywhere one looks, there are signs of the externalising of agricultural production, muted after the fall in commodity prices but still evident. Opportunities continue to be sought in parts of the world where productivity could be raised by introducing up-to-date methods. Some British farmers had invested in Hungary before it joined the EU in order to acquire cheap land and avoid regulation. They ignored the threat to the great bustard from ploughing up its habitat in the puszta biome. British farmers are active in Russia too. Once private ownership was permitted, a rush to buy neglected collective farms followed. A British company, with £ 20 million capital, acquired 35,000 hectares in the Tambov region of the Black Earth grain-belt.[xxi] One villager enquired whether the company was going to export the soil to England — shades of the Nazis who back-loaded soil in boxcars from the Ukraine to Berlin! Meanwhile, South African farmers were leaving for countries elsewhere in Africa in order to 'invigorate' their agricultures, and Georgia in the Former Soviet Union was advertising for South African expertise.

THE CASE OF BRAZIL

The most stunning success in raising agricultural productivity has been that of Brazil, which is remarkable because the development is taking place in a region hitherto written off as hopeless. A still more encouraging prospect is that output could be raised much further if facilities like those of the American Mid-West could be replicated, mainly meaning better communications.[xxii] The Mato Grosso has been described as Iowa with bad roads.

Of the Brazilian crop, 70% is carried by road but the roads are poorly paved where they are paved at all. In 2010, it cost on average $103 per tonne to transport soya from central Brazil to ports in the southeast of the country, as against $22 per tonne from field to port in the United States and only $17 in Argentina.

This suggests that infrastructural investment and an extension of existing methods could unlock the full potential of the Cerrado section of the Mato Grosso. Brazil's interior is dominated by a high plateau originally covered in cerrado, a scrub similar to Californian chaparral. This area was dismissed as a wasteland but its soils have been made fabulously productive. Despite its infrastructural drawbacks, Brazil has become the world's largest exporter of beef, chicken, orange juice, green coffee, sugar, ethanol, tobacco, and soya beans, meal and oil. It is the fourth biggest exporter of maize and pork and has a herd of 205 million beef cattle, which is the second biggest herd of cattle after India. Even so, Brazil has only one cow per hectare, a density thought to be capable of doubling.

Brazil's agricultural productivity is rising by 2% per annum: grain output has risen 152% in 20 years on a planted area that has expanded by only 25%. The country has become the first tropical agricultural giant, with all this implies for warmth in the growing season as well as for length of season, and is the first tropical country to challenge the big five food exporters, United States, Canada, Australia, Argentina and the EU. All told, the country probably has the world's largest agricultural trade surplus — without paying state subsidies. It has led the emerging economies in challenging the rich-country tariffs that stifle productive potential in a world that has to feed seven billion people and may have to feed nine billion by 2050.

The Cerrado produces 70% of the country's output. Farmers from the south began to trickle in during the 1960s and international agri-businesses were quick to follow. Holdings are huge and grow GM crops. The land is flat, relatively cheap, and produces three crops per year. High acidity was what led the area to be ignored for so long, until it was realised that spreading the underlying chalk, i.e., applying lime in industrial quantities, was what was needed. Applied science has really paid its way here, thanks to Embrapa, the Brazilian Agricultural Research Corporation, which is the leading tropical research institute in the world. Among other developments, Embrapa has found ways to fix soil nitrogen; bred grass highly suitable for cattle feed from *Brachiaria*, an African grass; and developed varieties of soya suitable for tropical conditions. 'Knowledge applied to nature' is its watchword. It has not escaped Brazilian attention that, contrary to conventional thought, agriculture and natural resources may become leading economic sectors. A partial analogy is Finland's conversion of the wood-pulp company, Nokia, into an international mobile phone giant.[xxiii]

Further expansion will mean clearing additional tracts of the Cerrado. This occupies one-quarter of the area of Brazil and has super-rich natural ecosystems, although they are less well known than the Amazonian rain forests. The area holds more plant and animal species than any other savannah in the world, yet has the lowest number of protected areas in the country. To the west of the plateau lies the Pantanal, the biggest wetland in the world, equivalent in area to France or Florida. A highway across the Pantanal stopped half-built or it would have skewed the drainage pattern and destroyed the wetland. The Cerrado is the most threatened biome in Latin America and its fauna are little studied.[xxiv] It is on the Cerrado, rather than in the Amazon rainforest, that the most extensive ecological changes have occurred. In 35 years, over half of the original 2 million square kilometres has been converted to farming. Further expansion is clearly on the minds of influential agronomists and big environmental losses may be the price. The area is losing ground faster than the rainforest and at the present rate of conversion may be gone by 2030.

The technical changes that have taken place entitle Brazil to claim an 'agricultural revolution'. In the type case of England, the concept of agricultural revolution has been discredited, through being applied indiscriminately to overlapping phases of rural transition, but the sheer speed of change in Brazil surely rescues the term. In a headlong rush, range land has been occupied under insecure titles and the area gives off a distinct scent of lawlessness. As a result, the prospects for conservation are not encouraging. The Minister of Agriculture stated ambitiously in 2011 that there are 120 million hectares of 'degraded' land which could be converted into farmland, tripling the cropped area.[xxv] A more cautious estimate is that the remaining area of so-called degraded land is closer to 20 million hectares, although even this might be enough to meet the increase in world food demand over the next decade. Former eucalyptus and pine plantations have been inserted into what *The Economist* describes as 'a litter of tree stumps and scrub', though charcoal burners have sometimes reduced the root-balls to fuel. Other fields have been levelled and made ready with lime and fertiliser and yet others have already been turned into 'white oceans of cotton'.[xxvi] In a *Financial Times* supplement on 'Investing in Brazil', the headline 'Farming is expanding without destroying the environment' seems aggressively optimistic.

Is the Brazilian agricultural model exportable? The region of obvious need is tropical Africa, from which Embrapa took the *Brachiaria* grass to improve pasturage (it upgraded Zebu cattle from India to feed it off). Brazil demonstrates that a big technological backlog is awaiting exploitation in Africa: for example, only 3% of the maize seed used in Ghana, which has excellent soils for the crop, are hybrid, against 90% in Brazil. Unlike developed country research stations, Embrapa has expertise in growing cassava and sorghum, which are African staples. And besides the dry Cerrado, Embrapa has worked in the wetlands along Brazil's Paraguayan and Bolivian borders. As *The Economist* says, in a remark more likely to cheer the world's hungry than it will please environmentalists, Africa 'needs to make better use of similar lands'.

PROSPECTS FOR WORLD AGRICULTURE

Expectations concerning future trends in productivity — not merely whether it will improve but whether it will do so faster than population growth — are split between gloom and hope. Spot observations surely encourage the latter, because plenty of agricultural experiments are being made on lab benches and the results are still to enter commercial pipelines.[xxvii]

Selecting hopeful signs over a very diverse range and almost at random: the Rinderpest virus was eradicated globally in 2010; plants have been developed with genetic material from a single parent, the easier to pass on desirable traits; sequencing the first plant genome took seven years and cost $500 million but the process is predicted soon to take only two to three minutes (sic) and cost a mere $99; resistance to GM crops, though still implacable in Europe, is waning in less-developed countries and GM now accounts for 10% of crop land globally. To continue, major finds of phosphorus deposits have been made; Dutch scientists have cropped salt-soil potatoes developed by traditional cross-breeding; the Norwegian government has opened the Svalbard Global Seed Vault; crop insurance is being extended among the peasant farmers of the less-developed world to reduce the instability of their incomes, while great improvements are being made in their business management as a result of mobile telephony and information networks.

Nothing dictates that these innovations, and the many others which might be cited, must continue producing sufficient gains in yield to overtake population growth. The temptation is plain for the media to jump the gun in reporting advances like these, just as it is to make premature announcements of cures for cancer. Nevertheless there are grounds for hope. The high prices of 2008 to 2010 recapitalised farms in the Midwest and elsewhere, offering an odds-on chance that output will hold up during future slumps. If high prices continue, agricultural R&D expenditures are likely to recover. Research is a lagged function of perceived need. Interest was relatively high during the 1950s, 1960s and 1970s, but sank along with the complacency induced by lower prices in the 1980s and 1990s, following the successes of the Green Revolution. At that stage, the attention of governments and the public, unhappy about the excessive use of agri chemicals, shifted to the environment. New chemicals did continue to enter the market but in the first ten years of the twenty-first century the rate of introductions fell by half.

At the end of that decade, when breaths were being drawn after the price spikes ended, the mantra concerning the future of farming became 'sustainable intensification.' Output is somehow to be raised enormously without an adverse effect on biodiversity. Everyone parrots this, but how it is to be achieved rarely appears. The head of the European Environment Agency has stated that the shrinkage of agriculture in Europe will mean declining biodiversity.[xxviii] Rather than approving of agriculture's retreat from marginal areas in Europe, or expecting this to favour wildlife, she thinks the effect will be negative: feral animals will form a reservoir of animal diseases because there will be fewer farmers to control them.

Implicitly taking a contrary view, the zoologist Matt Ridley has made some optimistic calculations. His assumptions are more conservative than those of the older, strikingly bullish, economist, Colin Clark, who extolled humanity's ability to feed itself.[xxix] Ridley claims that, with an entirely possible increase in yields, the world could feed itself on existing farmland or even less. He notes that in Western Europe and Eastern Asia, the application of fertilisers to the farmed area so raises yields that people consume only half of all plant production, leaving ample for other species. His account of the thriving wildlife on arable land in England and Western Europe nevertheless strikes me as far-fetched, based on my observations of birdlife in southern England, begun a decade before Dr Ridley was born! It is well established that the numbers of breeding birds on farmland are far fewer than only a few decades ago, when agriculture was less intensive, and visiting species are present in any given field for shorter spells.

Raising output worldwide without taking in new areas seems unlikely, even out of the question. Politicians will prefer this as an easier solution than awaiting and delivering research findings to raise output per hectare. Doubtless more food will be produced per hectare — China has doubled spending on agricultural research in five years — but across the world an intake of virgin land seems unavoidable. From the ecological point of view, this is more damaging than raising yields. The expected ratio between yield increases and additional acreage is mooted as 2:3 although, like the interminable get-out clause of fund managers, 'past performance is no guarantee of future performance'.

Innovation alone, diffusing best practices rather than inventing new ones, could raise productivity enormously. Brazil shows the potential gains from adopting the methods of the Midwest. Still more are possible if Brazil can close the remaining gap, generalising the use of bio-engineered seeds and the computer monitored planting that has given the Midwest two rows where one was sown before. Compared with most of the less-developed world, Brazil has special advantages. The Cerrados, where many of its advances have been made, started as *tabula rasa;* most agricultural systems in the less-developed countries are not so favoured. They are composed of tiny peasant plots where economies of scale could be obtained only by throwing large numbers of holdings together, obliging independent farmers to work as wage labourers or leave for the cities. Many people are eager to leave the land but financing enough urban infrastructure, housing and employment for them is challenging. Changes in land tenure are politically fraught.

One of the most sensitive places is China, where opening up the land market and permitting enough sales to raise the average size of holding might — the authorities fear — threaten public order. The conclusion might have to be that the diffusion of advanced methods will continue to be slowed by tenurial and political limitations. If this is so, the further conclusion may be that extending cultivation into previously under-used land is the practical alternative, whatever it entails for biodiversity.

The faster technology advances and the more best-practice diffuses, the more limited the taking in of wild lands need be. Optimism is grounded in experience but is a non-refutable postulate, so one author claims, because it is based on faith in technological progress and on the 'neoclassical' belief that man-made capital can be substituted for natural capital.[xxx] So-called 'strong sustainability' contests this by urging that the loss of natural capital is often irreversible; some natural capital is essential for basic life support;

and because people are averse to its loss, they do not believe that greater consumption compensates.[xxxi]

Arguments like these contain a mischievous element. Everyone knows that forecasts, i.e., propositions about the unobserved future, cannot be refuted. But what is to be done in practice? There is no humane alternative to trying to raise farm output. Eliminating natural capital to the point where human existence is threatened would of course be self-defeating; great care is needed because places do exist (the Aral Sea, for instance) where devastation really has reached the extreme. Whether a majority of people are genuinely disinclined to trade natural capital for consumption is nevertheless doubtful. The 'Crete Scenario' seems likely to satisfy much of humanity, especially the billions who are desperately short of food: they might accept an artificial landscape provided they are fed.

Rather than pursue an inconclusive philosophical line, it might be better to canvass scientific opinion about the prospects for raising productivity. One pertinent article observes, and other sources concur, that more food is currently produced than is needed to feed the world population.[xxxii] The problem lies less in expanding total output than in distributing the harvest, or entitlements to it, equitably. This difficulty is super-severe but it is a political matter rather than a question of agricultural capacity. Another paper claims that there will continue to be a steady stream of new plant and animal technologies.[xxxiii] This well-informed contribution does concede that private sector research is not a perfect substitute for the recent decline in public expenditure, but once again this is more a political issue than one of biological capacity.

The political colouring of the debate is obvious in a third paper which strains to show that alternative agricultural practices by small farmers would be capable of feeding the world.[xxxiv] It does have the merit of asking whether food security and biodiversity are compatible goals.[xxxv] Reading widely, it is hard not to be struck by the discrepancy between the alarms of the food security lobby, eagerly repeated by the press, and the confidence of scientists in continual long-term advances in productivity. For example, it took Professor Keith Edwards at Bristol almost two years to decode the wheat genome but the technology is now improving by several orders of magnitude annually.[xxxvi] The terms of his funding required all the data to be made universally available. Edwards notes that governments in the developed countries have finally realised agriculture's importance and started funding research appropriately. For a natural scientist to admit there are sufficient funds is a breakthrough in itself!

In another example, pigs in Britain can be fattened for the kill in 3 months whereas they take 12 months in China.[xxxvii] Chinese practice requires 6 kilogrammes of cereal-based feed for each of half-a-billion animals for an extra 260 days. 'Take that away', says the export director of the British Pig Association, 'and you can see how the genetics industry can do such a wonderful job for Planet Earth'.

I end on the optimistic side of the ledger and cite in support the 2012 Rockefeller University study that claims the land needed for crops worldwide is at its peak and will fall by a tenth over the next 50 years.[xxxviii] Crop productivity is rising by 2% per year as against a 1% rise in population. As people become richer, their calorie consumption levels off, suggesting that rising demand in poor countries may be offset by falling consumption in rich ones. Nor is meat consumption rising as fast as was feared; many Indians, for example, are vegetarians.

ENVIRONMENTAL CONSEQUENCES

On the other hand, most American farmers are said to be uninterested in protecting the environment because in 2012 it became so profitable to plant every available inch.[xxxix] Technologists look enthusiastically on zones where land can be brought into production, and display little concern for nature (perhaps their view is proper in the face of malnutrition and starvation among the world's poor). Louisiana engineers are draining the Yala swamp in Kenya to plant rice, describing it as 'worthless malarial wasteland'.[xl] The World Bank refers to the great arc of the Guinea savannah — 4 million square kilometres — from Senegal to Angola, as 'the world's last large reserve of under-used land.' Fully developed, its indigenous flora and fauna would be replaced by simpler, more uniform and agriculturally more productive ecosystems.

Changing technologies and price incentives mean ecosystems are never static. Agricultural disturbances are profound, with negative environmental effects on- and off-site. For example, on tropical farmland, the taxonomic diversity and population abundance of the macro-fauna is typically less than half that in primary forest, though the range of organisms in tropical pasture may actually increase.[xli] Agriculture implies planning the range of crop plants. Its unintended consequences include attracting alien species in unfamiliar proportions, while some native organisms are likely to respond over-energetically to new opportunities for feeding and roosting. Some are likely to soar to the status of pests.

As Edgar Anderson expressed the matter, 'no world crop originated in the area of its modern commercial importance... With many tropical crops the center of production is even in the other hemisphere from the center of origin... ' He added that, 'where a crop was domesticated there are the maximum number of pests and diseases which have been evolved to prey on that particular kind of plant... The farther you get from its center of origin the more of its pests can you hope to leave behind'.[xlii] Unfortunately for agriculture, what is gained is likely to be only breathing space and, given the environment's supreme potential for contingencies, any lull tends to breed complacency that may later place the farmer at a disadvantage.

The sequence of changes can be thought of as three anatopisms, i.e., mislocations. The primary anatopism is the initial technology transfer, which brings crop species for the first time to lands of new settlement. There is an initial gain in output and markets may encourage abandoning the old ecological insurance of growing a mixture of crops in favour of the monocultural production of some great staple or export crop.

The secondary anatopism is the arrival of pest species, often also from abroad. They may leave behind the predators and parasites which kept them in check in their former range. In addition, there may be damaging mutations of organisms already present in the new locality. Pests are not neutral to the scale of crop-growing and can increase disproportionately relative to the area under monoculture if it provides them with congenial habitat.

Reducing, let alone eradicating, pests is difficult and encourages a search for a further range of species as agents of biological control. They are the tertiary anatopism. Examples where biological control agents have run right out of control are well known, for instance, the explosive spread in Australia of the cane toad, originally imported to control the insect pests of Queensland sugar cane. Induced ecological change thus leads to a never-ending battle and requires vigilance pursued with more wariness than has usually been forthcoming. When the guard is dropped, old pests and animal diseases, not to mention new ones, seize their opportunity. This happened in Africa during the Second World War: research and extension services went into abeyance and the agronomic gains of the 1930s were lost as plant and animal diseases spread into new areas.[xliii] Similar dangers are threatened today by political conflict and failed states.

The degree to which losses of wild plants and animals will continue and how far this matters remain contentious; what is to be done matters more. Extreme doomsters damage their case by exaggeration. They overlook the long evolutionary history of speciation as organisms adapted ever more closely to their environments, including habitats disturbed by human action. On a shorter historical timescale, they overlook the spreads and invasions induced by humanity; surprisingly, the critics take the crypto-conservative view that whatever is in place today, however artificial its origin, should not be tampered with in future. Faced with the need to feed the world, heavy losses of species and habitats are to be anticipated: 'agriculture always has a detrimental effect on the natural ecology — that, in a sense, is its purpose.'[xliv]

ENDNOTES

[i] Letter in *Financial Times*, 1 February 2011.

[ii] Supposed identities with the Industrial Revolution or the Fall of Rome are the two great shibboleths and typically strike the scholar as specious. See E. L. Jones, *Growth Recurring* (Oxford: Clarendon Press, 1988), p. 182.

[iii] *Financial Times*, 9 May 2011.

[iv] M. A. Aizen and L. D. Harder, 'The global stock of domesticated honey bees is growing slower than agricultural demand for pollination,' *Current Biology*, **19** (11) (2009).

[v] B. Kidd, *The Control of the Tropics* (New York: The Macmillan Company, 1898), p. 42.

[vi] *Financial Times*, 22 June 2011.

[vii] *Financial Times*, 13 November 2009.

[viii] *Financial Times Magazine*, 15 October 2011.

[ix] *Economist Technology Quarterly*, 11 December 2010.

[x] E. Jones, *The Record of Global Economic Development* (Cheltenham: Edward Elgar, 2002), Chapter 5.

[xi] *The Economist*, 5 February 2011.

[xii] *New Scientist*, 27 August 2011.

[xiii] *FDI Weekly Strategic Report*, 21 September 2011.

[xiv] D. S. Sheperd, E. Setren and D. Cooper, *Hunger in America* (Center for American Progress, October 2011).

[xv] E. L. Jones, *The European Miracle* (Cambridge: Cambridge University Press, 3rd edn., 2003), Chapter 4.

[xvi] *Future Directions International Strategic Analysis Paper*, 'Global Land Limitation,' 6 April 2011. Future Directions International, based in Perth, Australia, of which I am an Associate, is a prime source for information on trends in world resource use which has been drawn on here. See also E. Jones, 'The challenge of feeding East Asia,' *FDI Strategic Analysis Paper*, 18 June 2010.

[xvii] *The Economist*, 7 May 2011.

[xviii] I am grateful to Professor Minoru Yasumoto of Komazawa University, Tokyo, for providing information.

[xix] *Atlantic Monthly*, May 2010.

[xx] *Financial Times*, Business and Food Sustainability Special Report, 27 January 2010.

[xxi] *Financial Times*, 1 October 2010. On Hungary, see *The European*, 31 August 1998.

[xxii] *Financial Times*, Supplement, 'The New Brazil,' 29 June 2010.

[xxiii] *The Economist*, 28 August 2010.

[xxiv] L. C. P. Faria *et al.*, 'The birds of Fazenda Brejao: A conservation priority area of Cerrado in north-western Minas Gerais, Brazil,' SciELO, Brazil [Internet site], 2009.

[xxv] *Financial Times*, 7 March 2011.

[xxvi] *The Economist*, 28 August 2010.

[xxvii] See, especially, *Financial Times* Special Report on World Food, 14 October 2011, and also *The Economist* 'The 9 billion-people question: a special report on feeding the world,' 26 February 2011.

[xxviii] *Financial Times*, 12 September 2005.

[xxix] M. Ridley, *The Rational Optimist* (London: Fourth Estate, 2010), pp. 145–147.

[xxx] E. Neumayer, 'Scarce or abundant? The Economics of Natural Resource Availability,' *Journal of Economic Surveys*, **14** (3) (2000), pp. 307–335.

[xxxi] E. Neumayer, *Weak Versus Strong Sustainability: Exploring the Limits of Two Opposing Paradigms* (Cheltenham: Edward Elgar, 2nd edn., 2003).

[xxxii] P. Hazell and S. Wood, 'Drivers of change in global agriculture,' *Philosophical Transactions of the Royal Society B*, **363** (2008), pp. 495–515. See also J. L. Simon (ed.), *The State of Humanity* (Oxford: Blackwell, 1995), pp. 397–403, which emphasises nutritional vulnerabilities amidst adequacy.

[xxxiii] W. Huffman, 'Technology and innovation in world agriculture: Prospects for 2010–2019,' Iowa State University, Department of Economics, Working Paper 09007 (2009).

[xxxiv] M. Jahi Chappell and L. A. Lavalle, 'Food security and biodiversity: Can we have both? An agroecological analysis,' *Agriculture and Human Values*, **28** (2011), pp. 3–26.

[xxxv] Notably sober discussions of the issues are to be found in Chapters 8 and 9 of S. Briscoe and H. Aldersey-Williams, *Panicology* (London: Penguin, 2009). An excellent survey of agricultural considerations is Robert Paarlberg, *Food Politics* (Oxford: Oxford University Press, 2010).

[xxxvi] Lecture to Cirencester Science and Technology Society, 9 May 2012.

[xxxvii] *Financial Times*, 17 May 2012.

[xxxviii] Cited in the *Financial Times*, 22 December 2012.

[xxxix] *Economist*, 16 June 2012.

[xl] *New Scientist*, 23 June 2012.

[xli] P. A. Matson *et al.*, 'Agricultural intensification and ecosystem properties,' *Science*, **277** (25 July 1997), pp. 504–509.

[xlii] E. Anderson, *Plants, Man and Life* (Boston: Little, Brown and Co., 1952), p. 187.

[xliii] G. H. T. Kimble, *Tropical Africa: Vol. 1 Land and Livelihood* (New York: The Twentieth Century Fund, 1960), p. 147.

[xliv] S. Briscoe and H. Aldersey-Williams, *Panicology*, p. 234.

CONCLUSION

15. WHAT SHOULD WE CONSERVE?

> Too many natural scientists embrace the comforting assumption that nature can be studied, indeed should be studied, in isolation from the human world, with people as mere observers.
>
> *The Economist*, 28 May 2011

All nature is so full, said Gilbert White, that the district most examined produces the most variety. He can scarcely have meant that we should be content to see the natural world cut into fragments, however much interest may be squeezed out of backyard natural history or kitchen sink ornithology. Nevertheless, he reminds us that interest remains in the smallest plot. It should be equally obvious that the ecology of areas affected by human activity is entwined with their history: White wrote the *Antiquities* as well as the *Natural History of Selborne*. Ecosystems are seldom simon-pure, they are historically contingent.

In practice, biological history and cultural history too often keep to their own sides of the fence. The leading proponents of English landscape history in the second half of the twentieth century, Maurice Beresford, Herbert Finberg and William Hoskins, were uninterested in nature, as they all told me themselves. Their equivalents in ecology, such as David Lack, whom I also knew, reciprocated by not attending systematically to human history. Since the post-war heyday of these men, environmental history has expanded into an academic discipline in its own right, though strangely neglecting its English pioneers. American sources even declare, mysteriously and inaccurately, that the subject began in 1978. Partly this stems from nationalism, partly from ignoring scholars outside the universities and partly from more interest in political concerns than attending closely to natural history on the ground.

Knowledge has become so compartmentalised and the volume of specialist work so overwhelming that a retreat into departmental bastions may be understandable. But tunnel vision is unfortunate. Naturalists can see people interested in preserving the landscape for cultural and historical reasons as competitors instead of collaborators. This Balkanises campaigns for environmental protection that would benefit from combined approaches.

When biologists do show an interest in the past, they are mostly concerned with historical ecology, such as studies of plant succession. These introduce the time dimension but as sequences rather than discrete phases. There is little interest in ecological history defined as the impact of ecological change on human welfare or the reciprocal effects of human intervention in the environment. A total of 980 scientific papers have been written about 'Oxford's ecological laboratory', Wytham Woods, but no actual history of the woodland seems to have appeared since the mid-1950s.[i] Any general pronouncement must be subject to exceptions, but reading older generations of biologists suggests some had broader interests than their successors. They were trained in Classics at school. From the other side, mainstream historians seldom incorporate environmental issues and, when they acknowledge the existence of the natural world at all, seem to think of it as an inert background to human affairs.[ii]

The rise of environmental history may appear to have been designed as a bridge, but it is not so. It is a branch of history and makes its contribution mainly to that subject. Occasionally, it has shades of anti-capitalism and when deprecating modern development has further tinges, if only implicitly, of people-hating. As it now exists, environmental history is a child of the 1970s; there were impressive precursors but they had been unable to attract many followers. The United States was the site of some of the first work, at least as early as George Perkins Marsh in the nineteenth century. By contrast, about 1960, I could find few publications directly on the ecological history of England. In 1963, Colin Tubbs and I began a paper in *Nature* by stating that, 'little work which falls squarely and designedly under the head of ecological history has been carried out in Britain'.[iii] We did not anticipate being contradicted and we were not. Further British work followed slowly — Tubbs, *The New Forest: An Ecological History* (1968), being a pioneering volume — and the field is no longer quite as narrowly American as it once was.[iv] There are now European journals of environmental history.

Archaeology is more satisfying to natural scientists because it appeals to physical evidence and pays attention to periods long enough to hold out the hope that biological processes will shine through. Documented history has less appeal, despite standing in relation to archaeology rather as medical examination does to autopsy. This asymmetry — as usual there are exceptions but I am concerned with tendencies — means that formative influences on ecosystems are neglected. One consequence is mistaken chronology. Current ecosystems are commonly assumed to be self-sustaining and of long standing, whereas research can show them to be artefacts of land-uses operating in the not-so-distant past. The assumption betokens a one-sided attitude to landscape preservation in which historical remains are disregarded and with them the practices that created them. The dwindling of old methods that were responsible for many elements in the landscape is especially disappointing and examples will occupy much of this chapter.

Grove and Rackham noted that where older practices have ceased around the Mediterranean, the open spaces on which they took place are typically frittered away on low density housing, golf courses, pine plantations, quarries, rubbish tips and the like.[v] 'Frittered away' is prejudicial, as if human uses are of no account, an attitude prevalent among naturalists. Yet it is true that the earlier landscape has not been treated sensitively. Even where land is in the public domain, as some lowland heaths in England are, the failure to persevere with or to re-establish the procedures that sustained them (livestock grazing, or burning to encourage spring growth of *Molinia* grass) means that they turn into birch scrub, then into woods overrun by dog-walkers, whose local political influence is as strong as their pets are disturbing to wildlife. Cases in point are public open spaces in the Newbury district of Berkshire: Snelsmore Common has become woodland and Greenham Common has been turned into what some locals call a 'pooch park', with dogs roaming uncontrolled and the area made less hospitable to ground nesting birds than when Greenham was a controversial American air base.

Administrators, bureaucrats, councillors and politicians are responsible for these misfortunes but cannot be entirely blamed when there is little accord between historians and biologists. Outright conflict can occur between people set on conserving historical sites and those eager to develop them as reserves for nature. Tom Griffiths made a prime study of an Australian example.[vi] His article opened with an episode concerning a military base at Langwarrin, Victoria, which had been converted into a 'Flora and Fauna Reserve'. The conflict came into the open when a book about the history of the base could not be launched on the spot because, as a senior land manager put it, 'there's a very conscious policy not to acknowledge history.' The manager of Historic Places on Public Land in Victoria went further, saying that, 'the park management culture tends to eradicate the memory and relics of past European uses in favour of an image of naturalness and primitiveness'.

What was curious is that the aboriginal impact, increasingly brought to the fore in nationalist as well as professional circles, was also brushed aside. It was subordinated to a rhetoric of pristine wilderness designed to expunge Australia's European settler history but in the event dismissing all human presence. As Griffiths says, 'it is part of the "green" aesthetic to separate, and sometimes eliminate, culture from nature, and "wilderness" has become one of its simplest and most popular manifestations.' Such prejudice did not figure in the writings of the great early historian of the Australian environment, Sir Keith Hancock.[vii]

Alongside the anti-historical parade in Australia, or at any rate in Victoria, is virtually a form of ecological 'ethnic cleansing'. Trees deemed not to have been present before the contact period (when the whites arrived) are removed in the name of a purer, de-historicized, landscape. I have found a local government employee at the bottom of my garden in the outer suburbs of Melbourne making an inventory of trees to decide which should be removed. Non-indigenous species were a no-no, though without any acceptable justification for eliminating them. The choice of baseline is political. This is nationalism, not science.

Eventually, someone comes along to redress the man–land balance. Jim Sterba's *Nature Wars* offers a striking corrective to the laments of loss in the eastern United States.[viii] He reveals that the retreat of farming means tree cover has come back to occupy two-thirds of the original area of the Great Eastern Forest. With this has come an abundance of wildlife, including white-tailed deer, black bears, Canada geese and much else — too much for comfort in the suburbia which now occupies the region. Sterba describes the conflicts over culling the deer, instancing the Princeton Woods, New Jersey. As I found, explaining the need for a cull ran up against the sentimentality of urban members of the Princeton Institute. I have no doubt that Sterba is correct: human occupation is inadvertently 'farming' the panoply of wildlife that now crowds into the suburbs. A similar tale might be told about the English countryside, where badgers and roe deer are 'farmed', thanks to abundant maize strips (grown for game birds) and unmanaged woodland.

TRADITIONAL MANAGEMENT

Ecosystems resulting from human intervention are at a discount. The most direct assault on the conservation of traditionally managed habitats is a paper by Hambler and Speight.[ix] They do not recognise that land management may be intrinsically interesting; are dismissive of 'charismatic taxa' (i.e., those groups that people will pay to watch and enjoy); and deny a role for aesthetics in conservation. In their concern for invertebrates they perhaps share the characteristic attributed to the Almighty by J. B. S. Haldane, 'an inordinate fondness for beetles.' Certainly zoo keepers are trying to move from large mammals towards smaller endangered wildlife, including the cockroach which my five-year old granddaughter was induced to stroke in Bristol Zoo![x] Invertebrates are worth conserving for their own sake, but should this be at the expense of species which are more attractive to the paying public? Hambler and Speight do not acknowledge the rights of the taxpayer who is funding conservation and give the impression of disliking public access to reserves.

Tradition is hard to define, they state, which is correct enough, although one reason may be that documents about older methods of land management are seldom analysed. They declare that 'tradition' has not existed long enough for species to become adapted to it, presumably meaning uniquely dependent on the particular mode of managing resources. Conserving traditionally managed habitats, they urge, wrongly targets species-poor groups and those at the edge of their English range, meaning some of the butterflies that favour coppice glades. Moreover, they make the claim that much wildlife has survived despite, rather than because of, traditional management, and above all they insist that coppice becomes richer in species when it is neglected. Yet left alone coppice will return to woodland and the birds and butterflies of the glades will desert it. The resultant woodland may shield tiny invertebrates from sunlight but its attractiveness will be confined to those biologists who specialise in scouring the recesses of dead trees for microscopic organisms.

Hambler and Speight do not demonstrate that the dwindling of obscure invertebrate members threatens the viability of ecosystems. Theirs is an argument for conservational significance by number of species. The goal would seem to be maximum biodiversity and biomass (they do not provide an operational means of trading-off quality and quantity), achieved through an emphasis on invertebrates. Is this what the ordinary naturalist wants? When the authors say that rarity should be assessed in global terms, with no special consideration for species that are scarce because they are at the edge of their range in Britain, we might persist in asking whether this is what the British taxpayer wishes to finance?

Another of their claims is that restoring grazing on neglected grassland can let in seeds that grow into scrub, though observation shows that insufficient grazing more often leads to a takeover by scrub. They assert that grazing by sheep and the coppicing of woods, the two traditional practices that most exercise them, slow down plant succession. Why not? England was for centuries, probably millennia, farmed and managed to produce precisely that outcome and the remaining scraps of 'traditional' habitat reflect it. Coppiced woodland is cut regularly and held in a state of arrested succession in order to maximise the production of hazel wands to make hurdles. That is its point and it has in consequence a characteristic flora and fauna. Finally, Hambler and Speight hold that aesthetics is not worthy of scientific consideration.

For conservationists to pass up alliances with historical preservationists to save land from a ravaging by development seems wasteful. I put my views on this years ago in a letter to *The Times* ('Jungle Takeover', 7 September 1961), pointing out the prospect of joint action in response to a contribution by Max Nicolson, Director-General of the (then) Nature Conservancy. At issue was the management of chalk grassland, at that date coarsening through the removal of rabbit grazing after myxomatosis. I noted that reintroducing sheep flocks managed along the lines of various historical periods would interest historians and pre-historians of agriculture, historical geographers and historical ecologists. For years nothing happened, and although the Conservancy did eventually run a sheep flock on grass at Aston Rowant, Oxfordshire, the opportunity of simulating historical management regimes was missed. In 2012, work parties were still engaged in clearing scrub.

An 'Iron Age' farm has been reconstructed at Butser, Hampshire, but in England there are few equivalents for documented historical periods. These periods are better served overseas, for example, by historic farms in the United States. Even there they do not, as far as I can tell, run simulations of farming's effects on the vegetation, which I had partly in mind, but they do hold out the possibility. In England, as Bryn Green observes, plants and animals benefited from the long periods of depressed agriculture in the late-nineteenth- and early twentieth centuries. Nature gained from 'old kinds of farming very different from modern agriculture'.[xi] The depressions relaxed competitive pressures on old-fashioned practices, conducing to their survival and hence to the benign environments they fostered.

Green goes on to note the advantages of coppicing and rotational grazing. They simulate natural perturbations, thus 'creating richer ecosystems than the either completely unexploited or the overexploited ecosystems all too typical, respectively, of nature reserves and modern farming or forestry'. Green was writing in 1989, since when the combined ecological and cultural merits of these practices have attained just a little recognition. In 2006, the Common Land Major Project Manager for Natural England (the latest name for the Nature Conservancy) wrote a paper, 'to address some of the criticisms which have arisen in previous years, when it was felt that there was over-emphasis on the environmental qualities of Common Land, without hearing enough about their human and cultural values.'[xii]

Many reserves do now include notes on local history in their promotional literature, but history and ecology are no more successfully integrated than academic geography fuses its human and physical sides. Here and there old methods persist or have been restored — floated water meadows on the Salisbury Avon come to mind — but the heritage industry directs its energies at buildings, and treats landscape features as incidental. Remains of fishponds, dovecotes, badger-baiting pits, and an occasional pit for fattening swans, are scattered about like fossils from incompatible periods, suggestive but inert. The few archaic institutions for allocating rights over resources that are still functioning may be thought of as the equivalents of living animals and offer more inspiration than any fossil record. Yet they are not classed together as members of one group. In writings about land-use planning, mentions of older modes of management tend to pertain to single sites. Individual sites may be cherished but there is little effort to record, rescue or preserve the categories of practice that sustain the hangers-on among them. Traditional management is one of the most endangered species of all.

Until recently — until the present in poor rural parts of the world — humans exploited whatever wildlife was to hand. Their activities, once universal, are now residuals, carried on in fringe areas by the descendants of people for whom they were routine; hunter-gathering seldom pays well compared with regular work and is abandoned once markets expand. Where ancient and at times frankly barbaric activities continue on a large scale they usually pertain to luxury trades, such as the killing of ivory-bearing mammals to make ornaments for Asia or harpooning whales to satisfy 'traditional' Japanese tastes for whale meat. (Since Tokugawa Japan was closed off from the world between 1600 and 1868, a period when ocean-going vessels were banned, any whales eaten were the beach-washed few or those that could be herded into bays and killed from small boats. Few Japanese can have tasted whale meat, which was made popular only during the food shortage after the Second World War).

In lowland England, examples of the old ways already mentioned include the allocating of pasture for grazing and the coppicing of woodland for hurdles (latterly for garden fencing). More recondite are the gathering of gulls' eggs and the netting of salmon and elvers (young eels). These practices may be classified as relating to the production of food. There is a case for preserving some of them. When ancient practices do not threaten wildlife populations they might properly be retained out of historical interest and to preserve their distinctive habitats and assemblages of wildlife.

Civilized societies ought otherwise to reject the hunting, shooting, killing and maiming of birds and animals, however sanctified by usage. Although lead shot in cartridges is banned in Scandinavia and the Netherlands, its continued use in Britain is actually defended by the chairman of the British Association for Shooting and Conservation on the grounds of 'tradition'.[xiii] The release of millions of target game birds and tons of lead shot into the English countryside has serious ecological ramifications, as does the so-called 'management' of predators — a euphemism for killing them off. Much of this 'management' remains hidden from view on keepered estates and is selective 'pest' control by private fiat. Land ownership is massively unequal and much land is inaccessible to the public. This means that it cannot be known how far wildlife legislation is respected; it is impossible to police landed estates.

I am therefore emphatically not arguing for the 'traditional' killing of birds and animals for sport; it runs counter to two centuries of resistance to cruelty.[xiv] The costs of shooting wild creatures are starkest in southern Europe — Corfu, where Mirabel Osler found hunters with golden orioles hanging from their belts; or Malta, where the sparrow expert, Denis Summers-Smith, saw sportsmen turn their guns on butterflies when birds were scarce; and where a flock of greater flamingos was reportedly shot in October 2012. There are of course mixed examples: the huge numbers of skylarks and other small birds slaughtered in France and southern Europe are killed for food as well as sport. Shooting skylarks is officially licenced by the European Union, and French netters and shooters probably exceed the supposed limit of 600,000 to 900,000 birds per annum.[xv] The official French hunting body states that 1.2 million to 1.6 million are 'harvested' legally each year. Italian hunting interests, too, rely on tradition to justify their kill. Between 1994 and 2004, Italian police confiscated 180,000 birds, one third of them skylarks, from hunters returning from outside the EU. The kill is brought to Italy in refrigerated trucks to supply pre-Christmas meals.

Some killing grounds therefore lie outside the area where small birds were historically netted for Italian consumption; the activity has become international. The firearms are of improved patterns and nylon netting is a modern invention, as are the trucks and communications employed in the trade. 'Tradition' cannot be advanced in defence of this slaughter, even allowing for the argument that many of the birds that are 'cropped' (for free: hunters put in no seedtime before harvest) are 'surplus' and would not have survived to breed the next season. In France, that argument would not hold because the skylarks killed are migrants from central and northern Europe, where the species is in decline.

Accordingly, old means of exploitation are not justified here on the sole grounds that they are 'traditional'. Some traditions are noxious and, from the point of view of wildlife and arguably for reasons of social health, the sooner they are out of the way the better. Wildlife values ought to include the right of the public to watch birds and animals without private sporting interests destroying them before they can be enjoyed.

Some of these strictures may nevertheless seem to apply to practices which I would defend. One has to make a judgement rather than be constrained by the definitions and rules of bureaucrats and lawyers. Formal rules are always inflexible and have unintended consequences. In advancing what may seem to be exceptions to a general distaste for cruelty, I have three purposes in mind: conserving ecosystems (which divides into protecting species and habitats), preserving parcels of land devoted to specific historical processes and sustaining survivals of ceremonies and institutions. The last category is in the greatest danger.

Grounds for preservation thus include the intrinsic interest of a practice, its historical continuity, its value in maintaining habitats with unusual characteristics and the fact that once abandoned it may be more difficult to restore than the habitat itself. The existence of relevant historical documents would be a bonus. Practices worth preserving include some with aesthetic and educational value. The world would be duller were they abolished. They may prompt the observer to think more deeply about landscape history and inspire fresh thoughts about relationships between land management and biodiversity.

TRADITIONAL MANAGEMENT: A STOCK-TAKING

Let us now give body to the discussion by focusing on threats to the last 'old fashioned' harvesting of biological resources in the south of England. These practices and the habitats associated with them teeter on the brink of an extinction that would be deplored if it directly affected birds, animals or plants: they are aspects of cultural history and warrant the care and attention ordinarily devoted to conservation. Their loss does not stem from the outright aggressiveness of some Australian conservationists but might nevertheless be termed collateral damage from a-historical approaches to ecosystems.

Take, first, coppicing. There were always hazels growing up again, but for years the cut-over glades remained open enough to the light for butterflies that are not shade-tolerant. The decline of coppicing was very evident by the time of a survey in 1947 and the practice virtually petered out during the 1950s. A limited revival occurred for making garden screens, but these were out-competed by imports of hurdles back-loaded in trucks from Poland. Wildlife Trusts sometimes revive a little coppicing to open glades, which otherwise are so few that certain species of butterfly dwindle and have now become rare.

Whereas coppices are usually privately owned, some areas of pasture do remain in communal ownership. They are unimproved grassland or occasionally heather moor with a few grassy lawns, the communally managed New Forest being the largest example. Communal ownership without communal management has survived in some smaller places, which tend to be rented out and the proceeds divided among the owners of the rights. Genuinely ancient and still functioning systems survive in this attenuated guise at Malmesbury and Purton Stoke, Wiltshire and Luppit, Devon.[xvi] They are historically exceptional, the first descending from the time of King Athelstan, who gave King's Heath to the Free Burgesses of Malmesbury, and the second from the time of Charles I, who gave land to the poor of Purton Stoke as a token after enclosing the remainder of Braydon Forest. Luppit is a survivor in its own right.

As an example of the fragility of such arrangements, consider the Thames-side lot mead at Yarnton, Oxfordshire. Yarnton Mead had never been ploughed, seeded or fertilised and was a traditional, floristically rich, hay meadow of a type now exceedingly rare. It had been managed in much the same fashion from Saxon times or at any rate from the early Middle Ages. Rights to mow strips of hay were auctioned annually in a public house. The auction was followed by a ceremony in the Mead at which the lots were translated into marked strips by drawing from a bag a cherry wood ball inscribed with the name of a strip. This randomised the allocation and was a form of insurance that prevented any one person from ending up with all his or her strips in either a wet or dry part of the mead. Successful bidders then mowed the hay of their strips.

The ceremony continued into the 1970s, ceasing on the death of the last meadsman, Ed Harris of Paternoster Farm. No one came forward to take over his role. Despite the significance of the ceremony, which was a medieval practice and hence a type specimen of institutional history, the responsible agency (at that date the Nature Conservancy) made no arrangement for the practice to be continued. Historians from Oxford, three or four miles away, showed none of the interest that might have been expected and which might have persuaded the authorities to think twice about permitting the ceremony to fall into disuse. Through a failure of both bureaucratic and academic imagination, the ancient organisation of Yarnton lot mead came to an end.

Procedures for managing communal pasture persist in the New Forest. The establishment of Verderers and agisters still operates. The Achilles' Heel of the system lies in the fact that the number of commoners coming forward to exercise the rights attached to their holdings is falling, despite the remainder turning out more livestock than ever onto the Forest. Commoners find it more lucrative to sell their holdings to incomers, who are more interested in acquiring paddocks for ponies and riding horses than in playing a traditional role in the grazing regime of the Forest. The grazing regime is intimately connected with the habitats of several species of plants and birds for which the New Forest is a prime location.

Minchinhampton Common, Gloucestershire, is in some ways a healthier example of communally run pasturage.[xvii] It is managed by and for the occupants of adjacent holdings of 30 to 50 acres, to which rights of common attach. The land has never been ploughed or fertilized and the vegetation differs from nearby commons which were ploughed not long ago. Unlike the New Forest, there is no problem of falling numbers of commoners. No farmer is an incomer and some are the fifth generation on their farms. Where once all manner of livestock were run, and in the fifteenth century there was a sheepcote, the farming is now solely dairying.[xviii]

Even here it is too much to expect no changes to have taken place and the old Marking Day ceremony on May 15, when the cattle were turned out, came to an end in 2002 because of the foot-and-mouth epidemic. Yet the system of common pasturage remains fully functioning, overseen by an officer with the medieval title of Hayward. In many ways Minchinhampton Common is a more intact system than the much larger New Forest.

Turning from grazing to more direct arrangements for obtaining food, we may instance the collecting of black-headed gulls' eggs. In 2009, only 25 individuals held annual licences from Natural England to collect gulls' eggs, and only one-third of the licence holders were still active, all of them over retirement age. A licence is issued only when there is a 'traditional claim', which may imply that there will be no new entrants. The activity is probably being permitted to waste away. Unlike the Netherlands, officialdom seems to envisage no permanent protection on grounds of cultural history. On the face of it, the motives for restricting the number of licences are conservational, since black-headed gull numbers are falling. But population counts do not run back far and the fall is from a peak that was itself temporary. There is some suggestion that egg collecting may have become harmful because heavy metals concentrate in the clutches produced after the first ones have been collected, but the effect is slight and uncertain.[xix] It is not cited as a reason for restricting the licences.

Restrictions on collecting gulls' eggs raise the question of what baseline is in the minds of policy makers when they decide on the number of licences. The target population of gulls is unclear. Since any intervention may unintentionally elevate the fortunes of one species over others, forecasting is hazardous and the licencing decision is likely to be a matter of bureaucratic choice. The issue looms because what are taken to be 'natural' populations today are seldom anything of the sort. In long-settled countries, many birds have adapted, or partly adapted, to intensive land use. Some species rely on old practices such as heather burning to create good nesting sites — Dartford warblers on the New Forest heaths and (formerly) golden plover on the heaths of Holland and Germany. Their populations are anthropogenic in the sense that human intervention created their breeding grounds.

One motive formerly invoked for retaining traditional activities was to protect the livelihood of the poor who engaged in them. The Welfare State may have rendered this unnecessary; at any rate few people in small, rural places are willing to register as desirous of charity and step forward to take their share of the proceeds from renting out common land. Things were different in the past. In early nineteenth-century Yorkshire, four gangs of local farm labourers, called 'climmers', used to descend the cliffs at Flamborough Head and collect up to 1,400 seabirds' eggs per day to sell as food or for use in the manufacture of leather.[xx] But the 'sporting' guns of visitors so reduced the number of birds that, by the 1850s and 1860s, only two gangs found it worth operating. The climmers themselves had always left sections of cliff for the birds to nest. Theirs was a sustainable tradition and they were exempted from the provisions of the 1869 Seabirds Protection Act.

The men who collected eggs and birds on St Kilda and Sula Sgeir were likewise exempted from the provisions of the 1869 Act and the Wild Birds

Protection Acts of 1880 and 1881.[xxi] These sources of food were essential back-up for their communities in seasons when the fish shoals failed. Yet the Acts were not primarily what curtailed the tradition: economic expansion was already leaching it away, even if growth in the Hebrides often took the back-handed form of emigration.

Today, Ness on the Island of Lewis holds the last community in the British Isles that takes young gannets for meat, under the sole exemption granted from UK and EU law.[xxii] Every August, 10 men catch, kill and process 2,000 birds by traditional methods. This tradition is actually flourishing and the number of gannets is rising, which is why the RSPB has stayed neutral in the face of objections on the grounds of cruelty. Two wrongs do not make a right but the objections are anomalous when far larger numbers of game and other birds are shot every year, nor do objectors respect the cultural significance of the practice. Similar complaints of cruelty have been made against catching and eating puffins in the Faeroes and Iceland, but neither belongs to the EU and so escapes its regulations.

Protecting the livelihood of working men came into play in respect of elvers (glass eels or young eels) in the Severn fishery.[xxiii] In 1873, local landowners, citing definitions in an Act passed over three centuries earlier, contrived via the Severn Fisheries Board to assert the game laws and prohibit the taking of elvers. Several men from the poor dockland area of Gloucester were fined, and even put to the treadmill for offences that included fishing on the land of the presiding magistrate! The poor had no votes but nevertheless constituted an urban-industrial working community over which rural landowners had no ultimate control. After a public outcry, seasonal elver fishing was restored in 1876, though today the elver fishery is once more in decline.

Traditional Severn fishing for salmon with the lave net is also precarious, and in the early twenty-first century, the fishermen and the Environment Agency are in dispute. The Agency wishes to extend a ban on the lave net for ten years, by which time it is likely no young men will still be coming forward to take up the practice. Perhaps that is the intention. The recurrence of wrangling shows that dispute has become almost a tradition in itself. In the mid-nineteenth century, the men of Lydney had been granted permission by a local landowner to use the lave net during times of depression in the Lydney tinplate works and the concession was renewed by subsequent Lords of the Manor, whether there was unemployment or not.[xxiv] Given this quite exceptional encouragement, lave-net fishing for salmon and twaite (a shad) managed to continue at Lydney. In the mid-1950s, it was carried on by a dozen highly independent men who owned plots of land as a second source of income and whom I was able to watch fishing. One said he was part of a six-man concern which was a semi-family affair and tradition. The catch was sent by rail to Billingsgate and on average was lucrative.[xxv] Today netting is reduced to a minimal activity.

Netting, including the use of fixed nets (stake nets), has always attracted the ire of fly-fishing interests. The Tweed Acts of 1857 and 1859 outlawed certain netting techniques and reduced the season for nets on that river — while lengthening the season for rods! The Tweed was the first river to be put under the management of Commissioners, who comprised all those owners with 'fishings' of an annual value of £ 30 or a prescribed length of river bank. This neatly excluded the netters.

The population of Atlantic salmon is well known to have been in long-term decline, the chief reasons being river pollution and the building of weirs.[xxvi] This is true of the River Wye, which as far as salmon are concerned has had a 'desperate history'.[xxvii] Whereas immoderate netting at the point where the river debouches into the Severn estuary had reduced numbers to a low by the start of the twentieth century, the prohibition on netting in tidal waters from 1908 did lead to a great revival.

On the Towy in Carmarthenshire, salmon had been taken by coracle fishermen for many hundreds of years. Figures do not exist for most of that time but an article in the *Carmarthen Journal* for July, 2010, cites Great Western Railway data for 1883 — the fish were being sent to Billingsgate and elsewhere, about one-third being exported. An estimated 200,000 were netted that year; others were caught by rod fishermen. Salmon numbers rose during both world wars, showing that fishing did hold stocks below the river's carrying capacity. But coracle fishing was on the decline. Many men who had inherited rights to fish particular stretches of the water in their coracles sold out to fishing clubs and landowners during the inter-war Depression.[xxviii] In about 1980, only a few families retained their rights, continuing the traditional method of a net between two coracles. In 1999, coracle fishing was granted an exemption from prohibitions on using nets because it had 'a unique cultural and historical significance in Wales.'[xxix] But only 12 licences were issued and various additional restrictions were placed on the practice. In 2007, the Welsh Assembly cut the number of licences to 8 on the recommendation of the Environment Agency (in England), to protect sewin (sea-trout) and salmon stocks, even though a spokesperson for the Agency was still paying lip-service to the cultural significance of coracle fishing.

In 2009, the total catch was only 427 and Carmarthen coracle fishermen felt obliged to fight to protect their legacy.

It is impossible to avoid the conclusion that salmon fishing, like the taking of elvers, has been the source of many a skirmish in the class war. There is a parallel in the nineteenth-century assault on working-class pastimes involving the abuse of animals, like cock-fighting, while at the same time elite fox-hunting and the pheasant battue were exempted. As late as the 1970s, the numbers of people with other full-time jobs were accused by the Association of River Authorities of operating salmon nets, 'only at the expense of those who are wholly engaged in commercial fishing and to the detriment of other river interests.'[xxx]

No doubt many of these adventitious individuals did lack any interest in conservation but the implication was that people such as 'garage hands, and nurses' had no business fishing in their spare time. The same charge was not laid against fly-fishermen. Demands were made for the new regional water authorities to buy up commercial fishery rights, as the proprietors on the Wye had done in the early twentieth century. The chairman of one authority commented that this would however provoke public hostility. Yet no significant resistance has been prompted by a big programme of buy-outs in the early twenty-first century.

The angling interest is rich and organised, a few years ago raising £ 350,000 to buy the licences of net fishermen on the Dee estuary.[xxxi] Rod fishermen were purportedly catching only 10% of the salmon and said to be releasing 'many' of them, whereas netting took perhaps 20%. Given the uncertainties in these figures, eliminating one type of fishing rather than the other hardly seems justified.

The situation with respect to fixed nets (putchers) for salmon on the Severn estuary is particularly vexed. The activity is at its last gasp. In 2012, the Environment Agency capped the allowable catch of the only remaining full-time net fisherman below a level that would enable his business to continue, so he claimed when planning to take the Agency to court. In a remarkable acknowledgement of the cultural value of the practice, given that it appears to have been stifling it, the Agency issued a statement to the effect that, 'the net fishing in the Severn Estuary is unique, and has a (sic) important cultural and heritage value. To allow them to continue we have capped the number of salmon . . . '[xxxii]

Most traditional salmon fisheries had already shrunk. Lately, they acount for only the tiniest fraction of all salmon caught, and a derisory proportion of by-catches of species like twaite. The catch of twaite is negligible compared with the 50,000 juveniles caught each year in the screens of power stations in the Bristol Channel.[xxxiii] Nevertheless, an Icelandic vodka entrepreneur and 'avid sport fisherman' co-opted other rich anglers and public figures to suppress commercial fishing for salmon.[xxxiv] This, they argue, had 'decimated' salmon stocks. The claim is that their intention is conservational but if so it does not include conserving cultural history. The North Atlantic Salmon Fund has bought up 85% of all Atlantic salmon quotas, licences and permits, gathering the fishing in the hands of inordinately rich private fishermen. The procedure is a one-way street. Once traditional rights — legalistically and narrowly defined, just as rights were at the time of the Enclosure Acts — have been sold, they will never be restored. Common rights can be eliminated but scarcely reproduced. If one wants a fancy term for this, the market in bundles of common rights is anisetropic, working in one direction but not the other.

Consider, then, the fishery that operated since Saxon times at Mudeford, Christchurch, the mouth of the Avon and Stour (once Hampshire, now Dorset). At the end of Victoria's reign, this fishery was virtually a social institution made up of men living on the spot, who had an ancient prescriptive right to the fish of their own harbour.[xxxv] They netted salmon when the fish began to move upriver — hardly coincidentally to the fishing beats of upstream landowners. In 2012, the Environment Agency was involved in negotiating an agreement for buying out the rights. The only rights recognised were those of individuals who had purchased a licence in the previous two years — a period so arbitrary and brief that one would have only to turn one's back for the opportunity to have vanished for ever. A man might have been away serving in the forces for so short a time. Anyone without a very recent licence simply lost out. The Agency could surely have devised arrangements to preserve a modicum of traditional netting against some future recovery of stocks. It could have restricted fly-fishing, which is only a game, and maybe opened up the rivers for the fish to move more freely. But it is all over now, elite interests assisted by state bureaucracy have prevailed and the tradition has come to an end.

CONCLUSION

One bright spot in this negative litany is the finding that in mainland Europe cultural history and natural history have not been quite divorced. A number of conservation bureaucrats there retain some affection for older ways, witness their interest in transhumance and mountain hay meadows. In 2010 a conference in Romania was entitled, rather trendily, 'Mountain Hay Meadows — hotspots of biodiversity and traditional culture'. On the other hand, the situation in England remains dire.

Ecosystems on the whole are too intricate to be returned to supposed purity. It is impossible to reconstruct them totally, even supposing the concept has much meaning. There has been too much flux; syncretic systems are everywhere; inter-specific relationships are impossibly complicated and not all of them are deleterious. If they *are* deemed to be deleterious because some species have taken over from others, which mix is to be preferred? One policy to bolster variety in the landscape would be to retain ancient practices and the natural and cultural ensembles surrounding them.

Ample testimony appears in experimental economics to show that people dislike losses more than they value gains, and so it is in ecology. Dismay at the disappearance of species from plant or animal communities outweighs the pleasure and interest of newcomers. The arrivals are considered vulgar species spreading unwanted and upsetting the 'balance'. Yet there is nothing new about change and the colonists among organisms make the laboratory of the world the more intriguing.

The rejection of introduced species on the grounds that they outcompete natives ('natives' often being no more than earlier arrivals that have naturalised) is foolish as well as futile. Yet 'invasion biology' has established itself as a sub-discipline. Lists of introduced pests are produced by government agencies in many countries, only to be amended once an organism proves ineradicable and becomes part of wildlife familiar to the eyes of new generations. The entomologist, Hugh Newman, was more scientific when he urged that the existing fauna or flora represent cross sections of processes of change and are the proper subjects of study.[xxxvi] Why, he asked, should all causes be regarded as natural except the actions of the human species?

Another curiosity about conservation is the idiosyncratic selection of cuddly or dramatic species to receive the bulk of attention. Far from dispassionate science, conservation movements 'privilege' some species, the public demands it and will contribute to their rescue. After all, people do not go to zoos primarily to see invertebrates. Biology proper is not immune from this distortion; certain species receive disproportionate attention. Trimble and van Aarde examined publications on animals in southern Africa between 1994 and 2008 and found them heavily biased towards some groups and species, such as chimpanzees, leopard and lions.[xxxvii] Few studies were published that might have helped the conservation of other groups. Birds were intermediate in attracting attention.

Human occupation is carved on the earth as far back as it is possible to look and searching for the mirage of ecological virginity is futile. On the contrary, the intellectual interest of ecosystems modified by humanity — as they all are — should be plain. Ecological science has more than enough to do studying the natural world as it exists and would be the better for recognizing that it cannot sensibly refrain from incorporating ecological history. I return at the close to the wise words of Edgar Anderson on the neglect of the weeds hybridising on our doorstep and those of Hugh Newman when he noted that all fauna and flora represent cross sections of the processes of change.[xxxviii] These are the true and proper objects of study. Better to take the world as it is, subject only to calculated interventions to reduce the costs of crop pests, blockages to waterways and so forth. Let us appreciate the biodiversity that economic growth reveals or makes accessible, and study nature as it nestles in the historicized landscape.

ENDNOTES

[i] A. J. Grayson and E. W. Jones, *Notes on the History of Wytham Estate with Special Reference to the Woodlands* (Oxford: privately printed, n.d. [Preface dated 1955]).

[ii] This was why the comparatively broad-minded work of Fernand Braudel and the Annales School in France burst with such force onto the English-language scene in the 1970s.

[iii] E. L. Jones and C. R. Tubbs, 'Vegetation of sites of previous cultivation in the New Forest,' *Nature*, **198** (4884) (8 June 1963), p. 977. The work by Thomas was referred to in *Agricultural History Review*, **8** (1960), p. 57.

[iv] C. R. Tubbs, *The New Forest: An Ecological History* (Newton Abbot: David and Charles, 1968).

[v] A. T. Grove and O. Rackham, *The Nature of Mediterranean Europe: An Ecological History* (New Haven: Yale University Press, 2001), p. 384. For the case of the Brecklands in East Anglia, see the letter by E. L. Jones, *New Scientist*, 23 February 1991.

[vi] T. Griffiths, 'History and natural history: Conservation movements in conflict?' *Australian Historical Studies*, **24** (1991), pp. 16–32.

[vii] K. Hancock, *Discovering Monaro: A Study of Man's Impact on his Environment* (Cambridge: Cambridge University Press, 1972).

[viii] Review of J. Sterba, *Nature Wars*, in *New Scientist*, 1 December 2012.

[ix] C. Hambler and M. R. Speight, 'Biological conservation in Britain: Science replacing tradition,' *British Wildlife*, **6**(3) (1995), pp. 137–148.
[x] Madagascan Hissing Cockroach, a favourite of school parties.
[xi] B. H. Green, 'Conservation in cultural landscapes,' In D. Western and M. C. Pearl (eds.), *Conservation for the Twenty-first Century* (New York: Oxford University Press, 1989), p. 188. The merits of cultural landscapes are also emphasised in Bernd von Droste *et al.*, Cultural Landscapcs of Universal Value (Paris: UNESCO, 2009).
[xii] G. Bathe, 'Common land: historical and cultural reflections,' *Common Land Proceedings* (2006), pp. 1–5.
[xiii] Interview, BBC Radio 4, 11 October 2012.
[xiv] E. S. Turner, *All Heaven in a Rage* (London: Michael Joseph, 1964).
[xv] P. Donald, *The Skylark* (London: T. and A. D. Poyser, 2004), pp. 168–169.
[xvi] I am grateful for permission to attend the ceremonies to Sue Poole, Clerk to the Warden and Freemen of Malmesbury, Don Cakebread, chairman of the Poors Platt Charity, and to the late H. L. Haynes, secretary, for permission to attend a meeting of the Trustees of Luppit Common, Devon. Interestingly, Mrs Poole and Mr Haynes both read economic and social history at university.
[xvii] I am grateful for information from the hayward, Mark Dawkins, and the chairman of the graziers' committee, Peter Gardiner, during a Hayward's View of Minchinhampton Common in 2012.
[xviii] C. Dyer, 'Seasonal settlement in medieval Gloucestershire: Sheepcotes.' In H. S. A. Fox (ed.), *Seasonal Settlement* (Leicester: University of Leicester, 1996), p. 28.
[xix] K. Pickard, *Heavy Metal Pollution and Black-headed Gull (Larus ridibundus L.) Breeding Ecology*, Ph.D. Dissertation (University of Southampton Research Repository; ePrints Southampton, 2010).
[xx] J. Sheail, *Nature in Trust* (Glasgow: Blackie, 1976), p. 24.
[xxi] J. R. Baldwin, 'Sea bird fowling in Scotland and Faroe,' *Folk Life*, **12** (1974), p. 99.
[xxii] *BBC News*, Highlands and Islands, 25 August 2010; *BBC Scotland*, 19 January 2011.
[xxiii] W. Hunt, *The Victorian Elver Wars* (Cheltenham: Reardon Publishing, 2007).
[xxiv] J. N. Taylor, *Illustrated Guide to the Severn Fishery Collection* (Gloucester: Folk Museum, 2nd edn., 1953), p. 12.
[xxv] Author interview, Lydney, 23 April 1956.
[xxvi] A. Netboy, *Salmon* (London: Andre Deutsch, 1980), pp. 72–88.
[xxvii] W. M. Condry, *The Natural History of Wales* (London: Bloomsbury Books, 1990 edition), p. 76.
[xxviii] M. Green, *A Year in the Drink* (London: Fontana, 1982), pp. 121–122.

[xxix] 'Low-impact living initiative' website; 'South Wales' website, 9 May 2011.
[xxx] Netboy, *Salmon*, p. 98.
[xxxi] *Severn Salmon Online*, 18 June 2012; Victoria County History, *Cheshire*, **5** (2) (2005), pp. 104–114.
[xxxii] See www.itv.com itv West, 28 June 2012.
[xxxiii] P. S. Maitland and T. W. Hatton-Ellis, *Allis and Twaite Shad* (Peterborough: English Nature, 2003, Ecology Series No. 3).
[xxxiv] *Financial Times*, 12 May 2012.
[xxxv] C. J. Cornish, *Wild England of Today* (London: Seeley, 1895), pp. 48, 50. I am grateful for information to Mike Andrews, whose forebears were net fishermen at Mudeford and who himself pulled salmon nets as a boy.
[xxxvi] L. Hugh Newman, *Living with Butterflies* (London: John Baker, 1967), p. 203.
[xxxvii] M. J. Trimble and R. J. van Aarde, 'Species Inequality in Scientific Study,' *Conservation Biology*, **24** (2010), pp. 886–890.
[xxxviii] E. Anderson, *Plants, Man and Life* (Berkeley, CA: University of California Press, 1971); L. Hugh Newman, *Living with Butterflies*, p. 203.

GENERAL INDEX

PERSON INDEX

PRINCIPAL SPECIES

BIRDS

BUTTERFLIES